Smart Card Security and Applications

Second Edition

For a complete listing of the *Artech House Telecommunications Library,*
turn to the back of this book.

Smart Card Security and Applications

Second Edition

Mike Hendry

Artech House
Boston • London
www.artechhouse.com

Library of Congress Cataloging-in-Publication Data
Hendry, Mike.
 Smart card security and applications / Mike Hendry.—2nd ed.
   p.   cm.—(Artech House telecommunications library)
   Includes bibliographical references and index.
   ISBN 1-58053-156-3 (alk. paper)
   1. Smart cards—Security measures.   I. Title.   II. Series.

 TK7895.S62 .H46   2001
 006—dc21   2001018808

British Library Cataloguing in Publication Data
Hendry, Mike
 Smart card security and applications. — 2nd ed. — (Artech
 House telecommunications library)
 1. Smart cards   2. Smart cards—Security measures
 I. Title
 621.3'8928
 ISBN 1-58053-156-3

Cover design by Igor Valdman

© 2001 ARTECH HOUSE, INC.
685 Canton Street
Norwood, MA 02062

International Standard Book Number: 1-58053-156-3
Library of Congress Catalog Card Number: 2001018808

10 9 8 7 6 5 4 3 2 1

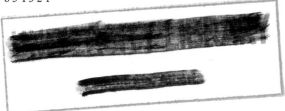

Contents

17   Personal Identification 231

18   Commercial Structures for Multiapplication Cards 239

Foreword

In the four years since the first edition of this book was published, one can say that nothing has changed and everything has changed. We have not seen new applications for smart cards, no real innovation: four years ago we were thinking of its uses in transport, telecommunications, digital television, and banking. We were discussing contactless as well as contact based cards. We were also describing the architecture of open architecture cards capable of carrying multiple applications. We were describing applications capable of being written for one chip being able to be run on different hardware types.

If there has been no real innovation it certainly cannot be said that there hasn't been considerable progress. What we have seen, however, is popularization. I don't think that it is an exaggeration to say that in the four years since the first publication of this book the smart card—the computer on the card—has come of age. It has become a mainstream technology. Most citizens in Western Europe at least will own one or two: a huge progression from the first edition when many people would never have seen one.

We have seen the acceptance of global standards: EMV and CEPS from my industry, finance, have become accepted alongside the established

GSM standard from the telecommunications industry. The bank card associations in Europe have defined a timetable for the conversion of all Visa and Europay/MasterCard branded cards to be chip based by the beginning of 2005. We have seen contactless becoming a production consumer technology with the success of the Octopus system in Hong Kong processing many millions of transactions annually. We have seen prices for multiapplication, open platform products drop below the magic $3 mark that could see them being issued as standard products ... and, perhaps most excitingly of all, the smart card has ceased to be a mainly European phenomenon: GSM has taken it over most of the world and in 2000 we started to see large-scale deployments of smart cards in North America.

The next four years can only see further progress.

Jon Prideaux, Executive Vice-President, Virtual Visa
Visa International, EU Division

Part I
Background

1

Introduction

The march of the card

Plastic cards are a part of the way of life in most industrialized countries. We use them to identify ourselves, to travel, to gain access to buildings, to obtain cash from our bank, and to pay for goods and services. We are regularly offered new types of cards; many people collect every card they are offered, while others feel that their lives are already excessively controlled by anonymous pieces of plastic.

Nor is this phenomenon restricted to rich western countries. Far Eastern consumers are among the most avid users of plastic payment cards, telephone cards, and travel cards; the card-based public-transport systems in Seoul and Hong Kong account for more transactions a day than all the world's electronic purses combined. In parts of Southern Africa, family benefits are paid in cash by a machine, by inserting a card and matching a fingerprint with that stored on record. Electronic purses are widely used in the former Soviet bloc and in Central Asia, where confidence in banking institutions is low, and preauthorized debit cards are used in Africa. Several cities in China use smart card public telephones, which are equivalent to the most modern in Europe or the United States, and the Chinese Central Bank has now also introduced electronic purse cards.

The first plastic cards were little more than a durable business card; the printed information made them more difficult to copy, but it was still read by human operators. *Embossing* made it possible to transcribe the information onto carbon-backed or chemical paper, but the data were still captured by key-punch operators for use in a computer system. *Magnetic stripes* enabled the whole process to be automated to a much greater extent, but the card was and is still essentially an identifier, at most linking the holder to an account. *Signatures* and *photographs* allow the identity of the cardholder to be compared, somewhat imperfectly, with that of the person to whom the card was issued.

Some of today's applications, such as health cards, retail loyalty, and portable data collection, require more data to be stored on the card than a magnetic-stripe card can comfortably handle. But the real reason for using a *chip* in a card, more often than not, is security: smart cards, and the integrated circuits used in them, have several features that allow them not only to store data in a secure way, but also to secure data stored in other computer systems. This book is concerned with these features and their use.

What is a chip card?

Chip cards are also often called smart cards or integrated circuit (IC) cards. Throughout this book, we will use the terms *chip card* and *smart card* interchangeably to denote a card meeting the ISO 7810 standard (bank card size and thickness), but incorporating one or more ICs within its thickness (see Figure 1.1). The term *IC card* is identical in meaning, but *chip card* is more easily related to the French *carte à puce* and the German *Chipkarte*. Many of these cards are in fact memory cards rather than microprocessor cards; both types are often referred to as smart cards, but some purists prefer to reserve the term *smart card* for microprocessor cards, whereas *chip card* can include memory cards.

Another term that can cause some confusion is *cardholder*. This does not refer to the person holding the card or even offering it at a point of sale; the cardholder is the person to whom a personal card was issued.

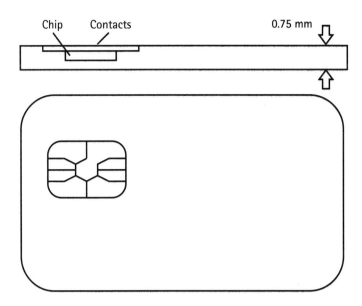

Figure 1.1 Chip card (or smart card).

Systems and procedures

Any security system is only as good as the systems and procedures that use it. Far more burglars exploit an open window or a weak door frame than pick the lock. Most computer system "hacks" involve a user ID and password, which either was divulged to the wrong person or should have been changed or canceled in the meantime.

Many security breaches involving cards do not exploit any flaw in the security of the card itself: Cards are lost or stolen, they are intercepted in the mail, or the fraudster arranges for the card to be delivered to the wrong house. Any proposals to increase the security of card systems must therefore address the whole system and not just the card itself. Using a highly secure cryptographic card badly is rather like fitting the lock to the wrong side of the door.

The security of modern cryptographic systems depends largely on key generation and key-management systems. To continue with the lock analogy, the security of a cylinder lock is not compromised because many people know how it works; but if someone makes a copy of the key or gives it to the wrong person, then the lock is useless. Cryptographic algorithms can be standardized and published, provided that appropriate systems are used to keep the keys secret.

Security is particularly important in chip card systems for two reasons:

♦ Many applications are in finance and payments, and potentially quite large sums of money can be involved. Up to 0.2% of the turnover in the major credit and debit card systems around the world—over $1 billion a year—is fraudulent. The main aim of using smart cards for these applications is to reduce fraud.

♦ Many chip card systems are used in sensitive areas such as personal identification and health. If security is compromised by inadequate systems, the resulting publicity could affect public confidence and reduce the scope for using cards—and indeed any computer storage of data—in these areas.

Market issues

The first chip cards were introduced in 1970 in Japan, and the idea was patented in Europe in 1974. This is scarcely a new technology, and it does suffer from having been promoted for many years without yet greatly affecting people's lives.

But the last few years have seen a dramatic turnaround in the fortunes of the smart card business. At the time of the first edition of this book in 1996, it was estimated that over 3 billion smart cards had been sold around the world. That number is now sold every year! Up to the mid-1990s, it could safely be said that no company had ever made money from the smart card business. This is no longer true: Gemplus, one of the leading French card manufacturers, in 1999 made a $35 million profit on sales of $817 million [1].

There are several reasons for this turnaround and these also give cause to expect further significant growth over the next few years.

♦ The credibility of the magnetic stripe has reached very low levels; consumers in most countries are aware of the major types of fraud (notably "skimming" and counterfeit cards) and those designing new schemes with any security implications are now almost certain to prefer chip cards to magnetic stripe. The main international payment card schemes have committed to putting chips on bank cards throughout Europe, Asia, and Australia. (There is at the time of writing no commitment for America.)

- The growth of commercial activity on the Internet—not only Internet shopping but also business-to-business activities—has given rise to a demand for better, more consistent security products usable on the Net. This application has given chip cards penetration of the North American market, which had previously been the main blank area on the chip card map.

- The runaway success of the GSM mobile telephony system, which places a chip card in every user's pocket, has not only driven market volumes, but has also pushed the development of cards with large memories, fast cryptography, and low power. The next generation of mobile telephony will bring the chip card to North America and Japan.

- Several other applications that are strongly linked to smart cards (such as health cards, customer relationship management, and satellite television decoding) are themselves growing rapidly in almost every country.

- The main patents have expired and many more players, ranging from small and innovative software companies to IBM, Sun, and Microsoft, are actively developing and promoting products. Microsoft's use of smart cards and certification in Windows 2000 is seen as particularly significant.

All of these are reasons for suppliers to invest in and be enthusiastic about the future of smart cards. They are also reasons why it is worth putting time and effort now into understanding the security mechanisms available, and into designing systems and procedures that will make full use of these mechanisms.

Organization of this book

The aim of this book is to enable readers to understand the technology of smart cards, particularly those features that relate to security, and the systems and procedures that must be put in place for a smart card scheme to be effective in protecting data. It should enable a reader to specify the security features required for a scheme in detail and, in many cases, to implement them directly.

This book is designed to be read by those responsible for the design, processes, or security of smart card systems in any application. Some readers will be highly technically aware and familiar with the concepts and terminology; others will be meeting them for the first time. Because some jargon and acronyms are inevitable, we will try to explain any technical terms when we first use them, and there is a glossary at the back of the book.

In the rest of Part I, we look at the background: the difficulty of defining the problem accurately when faced with a range of misconceptions, prejudices, and sometimes impenetrable computer and encryption jargon. Chapter 3 takes the reader through the process of analyzing and specifying security requirements and explains the range of tools available for risk management and requirements definition.

Part II explores the technology behind smart cards and related products, starting with the basics of card technology, including magnetic stripe and optical cards as well as IC cards. Chapter 5 is devoted to the science (many would say the art) of encryption and key management; it includes new sections on public key infrastructures and on the card authentication methods used by payment and GSM cards.

Identification of the cardholder is a critical problem for any card technology, and often a major role of the card; Chapter 6 looks at the different methods of identifying a person. Chapters 7 and 8 consider the different types of smart cards and the hardware and software elements that make the cards special. Many new features have been added to cards and systems since the first edition.

Next we look at the other system components, from card readers to computer networks, that can also play a role in enhancing or compromising the security of a card system. The need to protect the card at all stages of its life, from manufacture to disposal, is often overlooked in the early design stages; there must be a positive plan for all these stages, and this aspect is considered in Chapter 10.

Part III is concerned with the major applications of smart cards: in telecommunications, computer networking, finance and retail payments, health care, transportation, and personal identification. This part finishes with a chapter that looks at the special problems posed by multiapplication cards, particularly where the applications may be downloaded to the card after it is issued, or where the applications are administered by different parties. In Part III we also explore current trends and issues and recommend a methodology that can be used for the specification and design of any standard commercial system using smart cards to protect goods and data. Users with exceptionally high security requirements or with unusual usage conditions will find recommendations for designing to their requirements.

The appendixes include a list and brief description of the most important standards in this field, including card, encryption, and procedures standards. These standards are referred to repeatedly throughout the book. The book concludes with a glossary of terms and a bibliography of references and further reading on the cryptographic principles discussed.

Reference

[1]   Gemplus Annual Report, 1999.

2

Problem Definition

Perceptions

Those responsible for introducing plastic cards into our lives have always emphasized the security that they offer. For many years the public believed that cards protected buildings and bank accounts against fraudulent access and that the technology was only accessible to the banks and security companies that issued the cards and administered the schemes.

Up to a point, this was true; cards used for security and financial purposes incorporate very many different security measures, each of which adds to the difficulty of using a card fraudulently or producing a counterfeit card. Unfortunately, familiarity breeds neglect, and most of the measures so carefully put in place are regularly ignored by those who should be checking them. This has led to a great many abuses, which have provided a rich seam for both responsible and sensational journalism.

However, sensational coverage has outweighed responsible reporting, and public perception of the security of card systems may have swung from one extreme to the other. It has been claimed that any criminal armed with a $500 card encoder can manufacture cards that can be used freely in cash machines and retail terminals; this bald statement is simply not true. But the criminal can often get away with simple confidence tricks using such cards; there is therefore a measure of truth in the claim, and the efforts of card issuers to defend the infallibility of their systems have only served to make the lay reader more suspicious.

... and reality

The truth is that most abuses stem from breaches of the procedures and checks designed to protect the system. The losses resulting from cashiers in retail shops not checking signatures or verifying that the number on the card is the same as that appearing on the receipt is much greater than that from perfect counterfeit cards.

The abuses that occur because several relevant technologies are now freely available and understood by many competent engineers—both within and outside the criminal community—are smaller in number, but are generally regarded as more pernicious by card-scheme operators.

The first group of abuses (problems with procedures and checks) is often not malicious; the abuses are simply mistakes or shortcuts taken by operators who do not appreciate the consequences of their actions. Many of the checks we will describe in Chapter 4 require the operator or cashier to carry out a visual check or to exercise some judgment. These checks are often carried out in a cursory way or ignored completely: cards are handed back before the receipt is signed or the card face is never inspected at all. A system that leaves gaps or opportunities for this type of error is defective.

Abuses of card systems resulting from intrinsic weaknesses in the technology are, however, growing. As we will see, the most common card technologies are widely available. Many criminals and hackers have sufficient understanding of the systems and coding techniques involved to allow them to produce cards or card transactions that will defeat the weakest links in the chain.

Such attacks can be highly targeted. Some counterfeit cards could only be used in a vending machine or automated teller machine (ATM). Meanwhile, others might not work in an ATM but would pass the simpler checks performed by a point-of-sale terminal or the visual checks made by a cashier with a manual embosser (a "zip-zap"). Many popular forms of fraud involve an accomplice who can be relied upon not to carry out the check. By these means the fraudster produces a bogus financial transaction or gains access to a building or computer system.

Fraudsters have even produced fake ATMs; these look sufficiently realistic for customers to feed in their cards and personal identification numbers (PINs); the details are then captured and used to produce counterfeit cards with known PINs.

Counterfeit cards are most easily used in unattended situations where the appearance of the card is unimportant. International telephones used to be a popular target for this technique, but today it is Internet merchants who are most often affected. With magnetic-stripe cards, the card does not even have to exist at all—a card number is sufficient. (This is an area where chip cards are destined to play a major role.)

The overall cost of these frauds and abuses is much less than many people believe, though. Credit card fraud is about 0.08% of turnover, much less than the cost of most systems currently proposed to eliminate it (none of which would be completely effective). Benefit fraud with card-based systems is also under 0.5%, a small fraction of the fraud experienced by manual systems. Computer systems protected by cards and tokens are very rarely attacked, simply because there are many easier targets. Robert Maxwell, for instance, did not need to break any card security system to defraud his employees of billions of dollars.

Nevertheless, the systems are known to have loopholes, and the risk of a very large loss or a damaging physical security breach is rising continuously. A company that is the target of an attack draws no comfort from the 99 that have had no problems, and every case of fraud helps to undermine the system and public confidence in it.

Calculating the risks: probabilities and odds

Much publicity has been given to stories of factories in the Far East turning out thousands of perfect copies of bank cards and of computer programs that decode cable television signals being freely distributed over the Internet. These stories all have some basis of truth. However, bank card counterfeiting is still more difficult and less prevalent than currency (banknote) counterfeiting, which is estimated to have increased tenfold between 1993 and 1996 as the quality of color copiers increased. The levels of fraud and counterfeiting in the card systems are high enough to be worrying, but not high enough to cause the whole system to collapse; in many countries (including the United States) they are regarded as a normal part of the cost of business.

In 1999 an engineer named Serge Humpich successfully broke an authentication key used in French bank cards [1]. The authentication method had been in use since 1983, and the key itself was only 320 bits, which would be considered extremely short today for this application. The method used to break the key will be explained in Chapter 5, but the effect was to allow counterfeit cards to be produced, which did not necessarily belong to any valid account but would be accepted by the domestic payment systems provided that an on-line authorization was not carried out. In the reporting of this case (as a result of which the perpetrator was sent to prison for 10 months), it was easy not to notice that this type of counterfeiting is almost trivially easy for magnetic-stripe cards, but the banks' systems limit its effect.

The encryption systems used for analog satellite television, many of which have been broken, actually scramble the signal rather than encrypt it with an acknowledged strong algorithm; fast signal processing systems could in principle unscramble the signal without reference to the keys. Presumably this decision was made because it was believed that the extra cost of the set-top boxes for a strong encryption method was not justified. In today's digital systems, however, faster processors allow the use of strong encryption and these systems are unlikely to be broken for many years.

An important feature of any satisfactory security system is that it not only protects but can be proved to protect. As a minimum, the system must detect any intrusion or breach of security. For "strong" security, it must be possible to analyze every mode of failure and calculate its probability. The system designers must be prepared to stand up in court and defend the system against the evidence of expert witnesses.

No encryption system can guarantee absolute security; no matter how long or complex the key system, there is a minute chance that the first combination an intruder tries will be the correct one. Strong systems trust that would-be intruders also calculate the chances and do not waste their effort gambling on such long odds, but they also include monitoring systems designed to detect possible security breaches.

One of the most dangerous assumptions a system designer can make, though, is that all would-be intruders behave "rationally" (i.e., they follow the same thought processes as the designer). Such assumptions are often exploited by hackers and others who approach a problem from a different angle. Criminologists and psychologists can often describe thought processes that are very different from the norm but nonetheless have their own logic. It is often necessary to think laterally or to ask "what if?" There can be many more modes of failure than first meet the eye.

Technical communication obstacles

Another problem frequently encountered by system designers is the difficulty of explaining the problem to end users. One way to accomplish this is to seek a rational, preferably numerical, basis for the system design.

It is common for end users to demand "100% security" or "absolute guarantees" that the system cannot be breached. As we mentioned earlier, this is impossible with any scheme, and it is important to work to realistic criteria. It is often possible to talk in terms of incidents per 100 years or to establish a link between the cost of the security system and the cost of the worst incident that could occur. These concepts are understandable to both parties and often help to break down obstacles.

Computer salespeople often do not help their cause when they use abbreviations and acronyms. The glossary at the end of this book shows the large number of terms used in connection with data security and smart cards. These terms are often used loosely, and customers and salespeople may have slightly different understandings of their meanings. Customers are often unwilling to admit their ignorance of the terminology, and salespeople often repeat the terms used by technicians without fully understanding them.

For security systems in particular, it is important that the criteria are initially expressed in user terms. The translation of these criteria into computer terminology and technical standards should normally follow the first part of the system design.

In Chapter 3, we will consider how these criteria may be defined, analyzed, and verified.

Reference

[1]   Perfira, A., "Affaire Humpich: dix mois de prison avec sursis pour le pirate des cartes français," *Le Monde*, March 11, 2000.

3

Specifying the Requirements

The first stage in any system design is to specify the requirements. This applies to the security aspects of the system just as much as to its functions. In Chapter 2 we mentioned that it is important to specify the requirements in user terms rather than in computer jargon, and wherever possible these requirements should be quantified (expressed in numbers).

In this chapter we consider the aspects of security that we can control, the types and levels of risk a system may face, and the need to balance these against the cost of protection.

Security criteria

Security services

Security means different things in different contexts. The security required by a pair of lovers exchanging clandestine messages is very different from that required by the crew of a space mission. In an IT context, it is usual to divide security into five domains; systems addressing each of these are referred to as *security services*.

- *Confidentiality:* Ensuring that data are only accessible to those authorized to receive it. Lovers, generals, and businesspeople all often need confidentiality in their dealings. Sometimes even the existence of the data must be kept confidential; for example, a sudden growth in traffic between two competitors could indicate that takeover discussions are taking place.

- *Authentication:* Ensuring that each of the parties in an exchange is able to prove its identity to the other parties. Authentication can apply to objects (a PC or a smart card, for example) as well as to people.

- *Integrity:* Ensuring that the data (a message, for example) have not been altered since its origination.

- *Nonrepudiation:* Ensuring that the person or object who originated a transaction cannot subsequently claim not to have done so. Sometimes we also need confirmation of receipt.

- *Reliability:* This is sometimes omitted from the list of security services—for financial services transactions, for example, it seems to fall in a different category. But for IT security specialists designing nuclear power stations, reliability, integrity, and authentication are the most important services, and confidentiality may seem less important.

We can break these services down further, by relating them to the real-world needs that they address and the part that smart cards play in meeting them.

Safety

Computer systems that protect personal safety normally have a fail-to-safety provision, so that even if the whole system fails nobody is at risk. In some cases, such as *fly-by-wire* aircraft and space travel, this is not possible: Without the computer system, the aircraft is unstable and uncontrollable. In these cases, voting systems and plausibility checks are applied to all control outputs. It is impossible to eliminate every possible catastrophic failure mode, but the probabilities can be made extremely low—an application of risk management, which we will discuss later in this chapter.

Few smart card systems will be directly concerned with personal safety, but smart cards may give holders access to dangerous areas or permit them to do dangerous things. It is always important to consider first whether any such risks exist, because the "cost" associated with such a risk is very high.

Nondelivery

This is the risk that data in a communication system are lost. The cost of nondelivery often depends on whether we can detect the error. If we receive message number 133 followed by 135, then we can request a retransmission of message 134. If we do not know that it was missed, the cost might be an order or a transaction that we failed to transmit to the bank. In private and local networks, it is nearly always possible to detect nondelivery; in public networks, the risk of it passing undetected is much greater, and systems should build in message numbering or other checks.

Smart card systems are often concerned with recording transactions or events. It is easy to lose a single event unless it is numbered; numbering also helps check against duplication. Another form of crosscheck is double entry; we may, for example, both keep a tally of the total amount spent on each category of goods, and record the individual transaction amounts. Use of such checks will help detect errors, but we must still have some way of recovering from them and, if possible, reinstating the lost transaction.

Accuracy

Here we are concerned with the possibility that data may be recorded or transmitted incorrectly. This normally occurs through random errors (perhaps caused by a poor contact or electrical interference). Smart card systems must often have very wide electrical tolerance ranges to deal with the many different devices and contact conditions they may encounter; these increase the risk of errors passing through. In practice, the card itself, and its interface with the terminal, are rarely the source of these problems; they are much more often caused by the external systems. But the system design must take into account all possible sources of errors.

The cost of such an error will depend on the type of data: A single character error in a text string may be easily spotted and not important, while a one-digit error in a transaction amount, or a single-bit flag, may be much more difficult to detect and will have a bigger effect. Systems can make use of plausibility checks to help detect serious errors.

As well as the parity and cyclic redundancy checks (CRCs) that are widely used in computing and data communications, security-conscious smart card systems make use of message authentication checks (MACs) on messages and critical data fields. These techniques will be discussed in more detail in Chapter 5.

Data integrity

Data stored on a card or another computer system should be protected against alteration, whether malicious or accidental. This is an area in which some types of smart cards excel. Areas of memory can be protected against access by unauthorized applications or individuals, and some systems make use of memory that can be written only once, such as *write once, read many times* (WORM), or *flash memory*, which after being written can only be erased as a block.

Malicious attacks on the integrity of stored data are exceptionally serious when they involve encryption keys, expiration dates, or authorization codes. In these cases the whole purpose of the card system may be perverted. Designers should normally assume that *malicious attacks will occur*; the system must be capable of maintaining its integrity in the face of such attacks.

It is also important to ensure that transaction details are not altered during transmission to the host system. MACs, digital signatures, and other forms of encryption can again be employed and are more secure than other forms of checksum or CRC.

Confidentiality

Many smart card systems are concerned with protecting the confidentiality or privacy of information. This applies not only to the main data files stored on the card, which might include medical details or credit limits, but also to data stored on other computer systems that may be accessed by using the card, such as the balance on an account or the details of a company's customers, orders, or designs.

It is important to take a realistic view of the risk involved here: Many companies spend a fortune protecting the confidentiality of internal information and a similar fortune publicizing the same information! Even with slightly more sensitive information, the risk may be minimized by ensuring that only part of the information is stored on the card or transmitted at any one time; codes rather than text may be used to describe items. The risk should be quantified where possible: How much could the worst-case incident cost the company?

Some confidentiality requirements are imposed by law: in the case of payments or medical records, the cardholder is entitled to privacy, and the card issuer or scheme operator should ensure that the best current practice is followed. For small-value payments, the card may form part of the wall of privacy: Data protection officials in some countries have ruled that details of such payments (such as location or goods purchased) should not even be stored by the card issuer. In this case the card is the only place where they can be stored anonymously.

There is a big difference between malicious breaches of confidentiality (*eavesdropping*) and accidental ones (often called *leakage*). Leakage is usually caused by a system malfunction, whereas eavesdropping implies a weakness in the system design (a *systematic error*), which may be exploited over a period of time to maximize the damage caused.

Access controls and encryption are the tools most often used to *protect* confidentiality and privacy; it is usually more difficult to *detect* a breach of privacy.

Impersonation

Impersonation, or *masquerade*, is the risk that an unauthorized person (someone other than the cardholder) can make use of the functions allowed by the card. Many smart cards perform a personal identification role; in this case we are looking at the chance of others being able to pass themselves off as cardholders. We will consider this in more detail in Chapter 6, when we look at personal identification techniques, but from a risk point of view we must be able to assess and quantify the cost associated with a successful impersonation.

This is an area where it is particularly important to look at the whole system and not just at the card itself. Impersonation incidents can have a particularly damaging effect on public confidence in a system.

Repudiation

A scheme operator must often be able to prove that a particular transaction took place, that it was authorized in the correct way, and that it has not been altered subsequently. Neither the cardholder nor the merchant can then repudiate the transaction or claim that it never took place.

Digital signatures, using public key cryptography, are the answer to this requirement; they are often referred to as *transaction certificates* and may be stored by the scheme operator along with the transaction details. Smart cards can store the private keys used to generate these certificates.

The system designer must weigh the cost of including in the system an automatic method for handling demands for proof of transaction (verifying transaction certificates) against the likely frequency of such demands. The very existence of a stored transaction certificate will deter many spurious claims. Support functions such as this account for many of the hidden costs of any card-management system; smart card systems usually aim to automate as many of them as possible, but they cannot be eliminated completely.

Quantifying the threat

Possible outcomes and costs

We must now examine the system itself and list, in user terms, the possible outcomes of any of the security problems listed in the previous section.

For example, a breach of confidentiality could result in:

- The cost of an investigation and writing apologetic letters;
- The loss of a customer;
- Compensation;
- A lawsuit;
- Widespread loss of confidence in the system and loss of customers.

These are associated with increasing costs per incident, but probably also with decreasing likelihood. If the system has been designed to support customer service inquiries and incident investigation, then the damage can probably be minimized.

In a card payment system, nondelivery incidents could range from loss of a single transaction to a whole day's trade of one merchant. More widespread loss of data from a single incident can probably be discounted except in the context of a disaster scenario (such as the computer center being destroyed by fire). We said earlier that almost all nondelivery incidents can be detected, so the cost associated with this type of incident is that of recovering the data. Taking into account the average value of transactions, it could be decided that the cost of recovering a single transaction is higher than the likely loss—although the perception of system integrity must also be taken into account.

Where it is difficult to assign a monetary value to each of the outcomes, we can use a grading scale, for example:

- *Grade I:* Significant danger of death or injury;

- *Grade II:* Likely to endanger the survival of the company or business;

- *Grade III:* Likely to incur significant loss of business, external costs, or compensation;

- *Grade IV:* Likely to incur significant extra work or delays;

- *Grade V:* Nuisance value—small delays or extra processing requirements.

Objects threatened

Having considered the possible outcomes from the various types of security breaches, it is useful to categorize these according to the object that is being threatened. This also acts as a check to ensure that all of the outcomes have been considered.

Usually, the object most at risk is an item or set of data. This may be a single field in a personal or account record, a complete record, or the whole file. The most serious incidents involve key fields and access rights (computer hackers like to set up or use system administrator accounts). Fields or items that have broadly the same characteristics and are subject to the same risks can be grouped together.

Hardware can also be the subject of a threat: What if the card itself is lost, stolen, or damaged? A PC or disk drive that holds data can be damaged or stolen, or the modem that gives access to the data may malfunction.

Again, the most serious problems are when the hardware item forms part of the encryption mechanism. It is increasingly common to encrypt all files on a disk (for reasons of both space and security); many backup systems have problems with compressed and encrypted files, and whole volumes are sometimes lost—particularly where the backup was incomplete.

Causes and modes of failure

As well as listing the possible outcomes and estimating a cost associated with each, it is worth considering the ways in which each of these outcomes might arise. Specifying security requirements is usually an iterative process; the first time we do it we may have relatively little idea what the system architecture will look like or what its components will be. As the design progresses, it is easier to imagine ways in which the system could fail and what its weaknesses are. Trying to list causes of failure before the system has been designed, however, has the advantage of there being no preconceptions. And nobody can be offended because their design is being criticized! Security incidents are most often caused by:

- Hardware failures (whether permanent or transient);

- Software errors (usually caused by inadequate specification or testing);

- Network and communications problems;

- Procedural errors;

- Malicious attacks (often exploiting a failure in one of the previous categories).

Causes of failure can be categorized as:

- *Random:* Single failures that cannot be predicted, although it may be possible to assign a mean time between incidents (MTBI) or other probability measure to them.

- *Intermittent:* Incipient failures can often cause a large number of incidents and incur large costs. Logging and statistical analysis are the keys to detecting intermittent failures and must form part of the overall system design if they are to be useful when required. This is particularly necessary for the large distributed systems typical of smart card schemes.

- *Systematic:* Failures that will occur every time a particular combination of events or a particular sequence of transactions occurs. These are often system design or software faults, and they are common in new systems and software versions. They can be reduced (but never completely eliminated) by careful specification and testing.

- *Serial:* Where one failure is likely to lead to another. Maintenance and backup systems are often less secure than the normal systems, which is the opposite of the desirable situation.

- *Catastrophic:* Failure of the whole system or a large part of it. The reasons for this may be internal to the system (e.g., corruption of a database) or external (e.g., a fire or electrical failure). At this stage we are mostly concerned with security and cost implications; we need to establish the cost of a security failure in order to make decisions as to whether we need, for example, a second site rather than a backup database or an uninterruptible power supply (UPS). Disaster planning is a separate issue, concerned with keeping the whole business going.

Frequency of incidents

After examining the potential sources of failure and assigning a cost to them, the next stage when specifying the requirements is to specify a maximum frequency for each type of failure. Because of negative psychological effects on staff and customers, a maximum frequency should be given even for failures that have only a nuisance value.

The likelihood of failure can be expressed as a probability (e.g., 1% per year). But a more easily understood measure is usually the MTBI—this is the inverse of the probability, so 1% per year is an MTBI of 100 years. The result could look something like Table 3.1.

Risk management

We have seen that many card-based systems suffer from imperfect security. One reaction to this might be to insist on the best available technology in all areas. This approach is valid in some cases, but can result in an inappropriate use of resources. In the case of smart cards, it would result in

TABLE 3.1   Sample Requirements Specification

Failure Outcome	Possible Cause	Minimum MTBI (Years)	Per (Unit)
Customer unable to use card in shop	Card failure	3	Cardholder
Customer unable to use card in shop	Terminal failure	1	Terminal
Customer unable to use card in shop	System failure	1	System
Scheme unable to collect data from terminal	Terminal failure	1	Terminal
Scheme unable to collect data from terminal	System failure	1	System
Customer account error	Card error	100	Cardholder
Customer account error	Terminal error	100	Cardholder
Customer account error	Comms error	100	Cardholder
Customer account error	System error	10	System

military-grade cryptographic processors, very large segmented-memory requirements, and a card cost outside the pocket of most schemes.

The security measures must be matched to the application, and the way to do this is to repeat the process described earlier for the specification of the requirements. This can only be done after an outline design for the system has been completed.

The designer will compare the security-requirements specification with the actual likelihood of failure of the elements concerned. In the early stages of the design, this may only be an estimate; if a computer model is being used, then different values may be used to find which areas are particularly sensitive—it may be worth looking at these areas in more detail or specifying higher grade products.

For hardware items, MTBI figures are often available from the manufacturers (for any security-conscious system, they must be). They are usually obtained by a combination of calculation and testing. MTBIs are usually lower than mean times between failures (MTBFs) by a factor of two or more. For a large system, even quite long MTBIs may yield significant incident rates: A population of 100,000 cards with an MTBI of five years will yield over 50 incidents a day. Card life is discussed further in Chapter 8.

The result of this process is a table showing the probability value of each of the events considered (Table 3.2).

The first thing this does is to concentrate the mind on the areas that have relatively high probability values: These are areas where potential

TABLE 3.2   Probability Value Table

Failure Type	Cost	Probability (Per Unit Per Year)	Number of Units	Probability Value
Single card failure	$50	0.2	100,000	$1,000,000
Terminal failure	$500	0.3	1,000	$150,000
System failure	$10,000	1	1	$10,000
Random card error	$1,000	0.001	100,000	$100,000
Random terminal error	$5,000	0.05	1,000	$250,000
Card programming error	$5,000,000	0.001	1	$5,000
Terminal programming error	$250,000	0.01	1	$2,500
Operator private key discovered	$50,000,000	0.00001	1	$500

disasters loom and where additional effort is most likely to be well spent. In the case considered here, card failure is the largest single cost; increasing card life may be a very expensive proposition, but the operator could consider offering customers special wallets or insisting on a higher specification plastic.

The next stage is to look at the cost of possible countermeasures. A proper risk analysis should ensure that if a countermeasure reduces the probability value by more than the cost of the countermeasure, it will be adopted. In the case of a new hardware or software design, it is often difficult to determine in advance just how much the countermeasure will cost: We may study an area for weeks and conclude that there are no viable improvements. Judgments must be made.

When the process is complete, we have a list of the risks and probabilities that can be presented to senior management. The business may need to seek ways of protecting itself against some of the more serious effects.

Standards

Use of standards within specifications

Standards offer a shorthand way of specifying a group of requirements, and in some cases an independent laboratory can test equipment or software against the standard, providing a certificate of "type approval" that prevents users from having to check every aspect of the item's operation.

There is a very large number of standards applicable to chip cards in one way or another. Appendix A lists over 40 of them, and this is by no means a comprehensive list. Some of them define interfaces or procedures, while others set minimum requirements for physical or reliability aspects. In very few cases, however, is there provision for certifying products against the standard.

Standards are often a statement of best current practice, and in that context should be followed or specified wherever possible. Almost all commercial encryption packages follow the relevant standards very closely. Specifying long lists of standards to which a product must adhere is not necessarily helpful, however, and may obscure the more important requirements of the individual case.

In some cases, such as the ISO 7816 part 1 physical standard for smart cards, the standard is now regarded as an absolute minimum, and most manufacturers have set their own standards well in excess of this level.

Classes of security

There is a group of standards that grade security products and systems according to the level of security they aim to provide and set a mechanism for testing products against their stated aims. For some years these have consisted of:

- The Trusted Computer System Evaluation Criteria (TCSEC), used in the United States;

- The Information Technology Security Evaluation and Certification scheme (ITSEC), used in Europe.

The two standards are comparable but not identical. TCSEC grades products from D (the lowest level) to A1 (the highest) according to the level of security they are trying to achieve and the extent to which they are able to demonstrate that they have achieved it.

ITSEC gradings subdivide the security objectives into the headings:

- Identification/authentication;

- Access control;

- Accountability;

- Audit;

- Object reuse;

- Accuracy;

- Reliability of service;

- Data exchange.

A supplier may define a product's objectives by using any combination of these criteria, but common requirements are grouped together in predefined classes. For smart card systems, the classes F-DI (high standards of data integrity during data exchange), F-DC (high standards of confidentiality during data exchange), and F-DX (combines the requirements of F-DI and F-DC) are likely to be the most important.

Under ITSEC, products and their documentation are assessed by a commercial licensed evaluation facility (CLEF) using criteria for correctness and effectiveness; they are given a rating from E0 (inadequate assurance) to E6 (the highest level). For ratings of E4 and above, formal models and design processes must be used; some smart cards have reached level E6, but it is unlikely that a complete system could be certified at this level.

It is important to recognize that any of these ratings depends on the security target, or objective, claimed for the product: an E6 rating is of little value if the security objective is not the same as that required by the target system. Users should beware of seeking, or even using, products just because of their high ITSEC ratings.

Both ITSEC and TCSEC are now being superseded by the *Common Criteria for IT Security Evaluation*, also published as ISO 15408. The development of the Common Criteria (CC) was a joint project involving security organizations in the United States, Canada, France, Germany, the Netherlands, and the United Kingdom, as well as the International Organization for Standardization (ISO).

The CC use a combination of evaluation assurance levels (EAL) and strength of functions (SF), which can be mapped onto ITSEC or TCSEC. As with ITSEC, CC evaluation is carried out by licensed facilities on behalf of government bodies (the National Institute of Standards and Technology (NIST) and the National Security Agency (NSA) in the United States).

A set of protection profiles for a smart card IC has been developed by the main European chip manufacturers [1] and should now be used for future evaluation of such products using the CC. This protection profile covers two main features:

- The integrity and confidentiality of the data on the card;

- The integrity and confidentiality of the security-enabling features, in particular interprocess memory protection and memory management.

It does not include an evaluation of, for example, any authentication functions themselves—these would have to be evaluated separately.

Quality assurance

Increasing importance is being placed on quality assurance (QA) as a means of ensuring the correctness of an operation. Many government departments in particular will only buy from suppliers that meet the ISO 9001 or 9002 standards, or their national equivalents. ISO 9000 manufacturers face increased costs when they buy from noncertified suppliers, so the pressure continues downwards.

As with the functional standards mentioned earlier, the standards themselves should be followed insofar as they represent best current practice. The ISO 9000 standards enforce a high standard of record keeping, which is very important for a card issuing or customer help-desk operation. Many organizations, particularly smaller companies, have found, however, that obtaining and maintaining full certification against these standards represents an excessive overhead cost.

Documenting the specification

Initial system specification

The initial security specification for a system must include:

- A list of the security objectives (what is the system trying to achieve?);

- A maximum frequency for each of the outcomes considered, similar to Table 3.1.

This "bare bones" specification, which may be a single section within the overall functional specification, will form the basis of a system design.

Analysis and iteration

The design should then be analyzed and the risks associated with each element listed and quantified (as in Table 3.2). These can then be modeled with a simple spreadsheet—a process that we will come back to in Chapter 19—to find areas that are particularly sensitive or where expenditure could most easily be justified. Following redesign, the process is repeated until the overall system objectives are met and any weak areas have been covered.

There is a danger in relying too much on mathematical models or spreadsheets in a process such as this: It is easy to overlook a whole category of threat or to make unjustified assumptions about people's behavior. Any formal analysis should therefore be backed up with verbal explanation and a statement of how the system objectives are met. It is often helpful to describe how the protection mechanisms would work in the face of:

- Random errors, for example in communication systems;

- Hardware failures in any system component;

- Software errors;

- Malicious attempts to break the system, based on any of the criteria listed earlier in this chapter.

If the analysis holds for each of these cases, then it is likely that the structure of the system is sound. The descriptions should form part of the security specification. For a payment system or other system in the public domain, they should be carefully written, as they could be required if an attack reaches the courts and evidence of the system's soundness is needed.

Component security objectives

For a larger system, the process of specification must then be repeated for each of the components of the system. Where the top-level specification makes assumptions about the security of a component, these assumptions should form part of the security objectives for that component. If its designers find these unrealistic or unachievable, then the system analysis must be repeated with the new data.

Assumptions about the security offered by an individual component must also take into account the type of user involved, the application environment, and the way the component is being used in the system. The effectiveness of every tool and technique depends on its context. In Part II, we examine the tools available and the contexts in which they can be used.

Reference

[1]   See www.eurosmart.com.

Part II
Technology

4

Card Technology

Visual features

Cards have been used for identification purposes for centuries. Diners Club issued the first plastic cards (what we would now call a "travel and entertainment" card) in 1950; the first plastic credit cards were issued by Bank of America in 1960. These early cards were made of a single layer of polyvinyl chloride (PVC) printed with a very simple pattern. The customer's name and account number were printed on the card using raised lettering, or embossing, in gold or a contrast color, a process known as tipping. At the time, color printing was only generally available using rotary printing machines, which could not be used with the relatively rigid plastic cards. The addition of the embossing and tipping made sure that only plants built specially for this purpose could produce these cards, and because the cards were rare and considered to be of high value, people accepting them checked the signature carefully.

Today, thermal transfer printers made specially for printing bank-sized cards will fit on a desktop and can be bought for less than $5,000. Manual embossing machines are even cheaper. The original designs can be scanned and copied by any desktop publishing system. Card issuers therefore make use of a very fine printing process known as *microprinting* (see Figure 4.1) and complex patterns and color changes, which are designed to be distinctive and difficult to reproduce. Printing techniques developed for banknotes, such as *guilloches* (finely printed spirals) and rainbow printing (graduated color changes), are useful because they defeat the resolution of color photocopying or scanning. It is worth looking at some of the characteristics of a modern card, particularly a bank gold card or a pass from a security-conscious establishment, under a magnifying glass to see some of its special features.

In many retail situations, however, assistants do not have the time or the inclination to scrutinize cards this thoroughly (any more than they scrutinize bank notes). And there are too many designs in circulation for them to be able to know the distinguishing features of each. Holograms are used to provide a quickly assimilated check of the authenticity of cards belonging to one of the major payment schemes: Visa's dove and Master-Card's interlocking globes can be recognized quickly and are very difficult to reproduce accurately. On some cards, these same patterns are printed on the card with ink that is only visible under ultraviolet light.

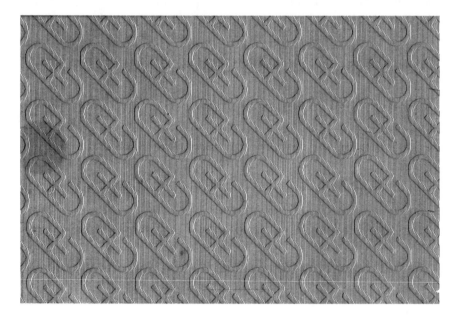

Figure 4.1 Microprinting. (Photograph courtesy of Giesecke & Devrient GmbH, Munich.)

Other visual features used on cards include the signature and/or the photograph of the cardholder. Either or both of these can be printed onto the card as a part of the personalization process rather than being added onto a paper signature strip. Here the difficulty is not only the thoroughness with which the signature or photograph is checked, but also that both characteristics often appear quite different in the small space allotted to them on a card from that in real life. Both signatures and appearance also change with time, and they depend on a highly subjective assessment on the part of an untrained observer. Something better is needed if the card is to be a reliable form of identification.

Magnetic stripe

The most common form of card technology for automatic reading is the magnetic stripe. A half-inch-wide (12.7-mm) strip of magnetic tape is bonded to the card substrate. The individual magnetic particles are aligned along the stripe, but when unencoded, the individual particles may be magnetized either from left to right or from right to left, so there is no overall polarization. Blank white cards with a magnetic stripe can be freely purchased from card manufacturers and dealers.

Copying and counterfeiting

Magnetic-stripe technology is essentially the same as that of a tape recorder. Most bank cards and identification cards conform to one single standard (ISO 7811—see Appendix A). This specifies the position of the data on the card, using three tracks, and the encoding scheme (known as F/2F). It is therefore relatively easy to construct a card reader from components, and in fact complete card readers with serial or keyboard interfaces are available from a wide variety of sources (see Figure 4.2).

Building an encoder is slightly more difficult, first because of the need to control speeds and tolerances quite accurately, but also because there are no standard components for conversion of data into the encoding current required to drive the head. However, it is a straightforward engineering task for a suitably equipped organization.

Because magnetic-stripe cards are used in such a wide range of applications, many organizations have a valid need to encode cards, and several encoders are commercially available at prices well below $1,000 (see Figure 4.3). Reputable encoders will usually reject requests to encode bank

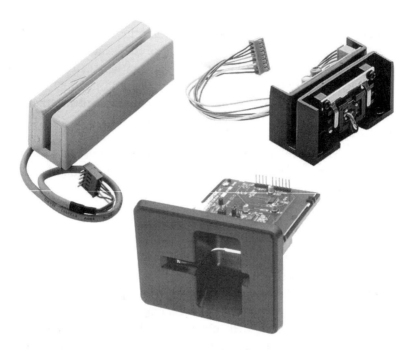

Figure 4.2 Magnetic-stripe card readers. (Photograph courtesy of Mag-Tek, Inc.)

Figure 4.3 Desktop card printer/encoder. (Photograph courtesy of Fargo Electronics, Inc.)

cards (unless this has been disabled; for example, when a machine is sold to a bank), but there are many other security applications of cards for which even this limited protection is not available. Designers of magnetic-stripe card schemes should always assume that cards can be reencoded or cards with identical encoding manufactured.

The technique known as *skimming* involves reading the data from a valid card and copying it onto another card. If the card is physically a copy of the original, then it may pass a visual inspection as well, but even white card copies may be used in many unattended applications.

Many security-conscious card schemes do incorporate techniques to reduce the risk of copying: Check digits and other algorithms can prevent new numbers from being generated, while ATMs and similar devices can leave tell-tale marks that are not copied by standard techniques on the card. By these and other means, as we will see in Chapter 14, the main bank card schemes have controlled the spread of fraud.

High-coercivity cards

The data on standard magnetic-stripe cards is often erased or corrupted by stray magnetic fields. This problem can be reduced considerably by using high-coercivity (HiCo) magnetic material. HiCo stripes can only be altered by applying a magnetic field stronger than most common permanent magnets.

HiCo cards also offer a higher level of security against copying. Encoding HiCo material requires more energy (higher recording currents), and it is difficult to ensure that data from one track does not spread into the adjoining track. Once encoded, a HiCo card is no more difficult to read than a LoCo card. But from a commercial security standpoint, the main advantage is that few commercial encoders can handle HiCo stripes.

The standard for HiCo encoding, ISO 7811–6, is not always used, so there is also scope for using different encoding schemes. In this case, someone wanting to encode cards would have to work out the encoding scheme as well as manufacturing a special device for the purpose.

Other magnetic card types

The magnetic stripe may be located in different positions on the card or it may be discontinuous. The magnetic material may be applied with a printing technique rather than as a tape, and this further increases the flexibility of the operator to locate data in different positions on the card.

These techniques will appeal to operators whose card schemes are large or critical enough to justify the expense of nonstandard card production, but which are not so widespread or offer such large returns that fraudsters will be tempted to go to similar lengths to duplicate the nonstandard cards. Japanese telephone cards, which have one side of the card covered in magnetic material, can be used as a basis for manufacturing counterfeit cards of this type.

Enhancing security using complementary technologies

Several methods can be used to improve the security of magnetic-stripe cards. Those that have been tried most extensively are:

- *Watermark tape:* A thin strip of magnetic tape, with the particles aligned across rather than along the tape, is bonded to the top of the card, on the same side as but above the magnetic tape. The watermark tape is read by special readers, which can be fitted to ATMs and in other critical locations. The tape contains a security code that is unique to every card; the code, or data algorithmically derived from it, is also contained in the magnetic stripe. If the stripe and watermark do not correspond, the card is judged to be counterfeit. This system depends on the proprietary nature of the tape manufacturing process; unfortunately, it also means that there is ultimately one source for the readers, whereas one of the attractions of magnetic stripe technology is the wide variety of reader makers. In trials by the major card schemes, reliability was also lower than would be needed for a full rollout.

- *Holomagnetics:* The concept here is similar; it involves using a machine-readable hologram in place of the visual hologram. Again, this is only suitable for special locations and types of reader; it could be used for protecting ATMs but not point-of-sale readers.

- *Card signatures:* Small variations in the quality of the magnetic tape or in the length of individual bits can be measured by specially equipped card readers. A "signature" made from these variations is stored, in encrypted form, on the standard tracks. These systems should detect not only counterfeit cards, but also unauthorized attempts to alter the data on the card. There are at least two proprietary systems (both from U.S. manufacturers) using these techniques, but they are not widely used in live systems. This is probably because magnetic stripes are always prone to changes caused by wear and magnetized heads in readers; such wear would lead to cards being falsely rejected.

Optical

Optical cards, using a laser reader detecting the depth of a reflecting surface on the card in the same way as a compact disc (CD), can be used to store a very large amount of data: the most common proprietary standard stores 6 Mbps on one side of a standard-sized card. The technology has been available since the mid-1980s. In the early 1990s, it was licensed by several manufacturers, which have developed commercial products suitable for reading the cards (see Figure 4.4).

The technology appears to be reliable, although because of the limited number of cards in use and the small pool of manufacturers, both cards and readers are still rather expensive. The quantity of data stored in non-volatile form on a card makes it attractive for applications such as health records and hospital data. The high cost and limited supply of readers ensure a reasonable level of protection for the data.

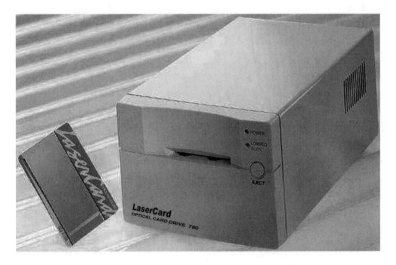

Figure 4.4 Optical card and reader. (Photograph courtesy of Laserscan Systems Corp.)

Despite these advantages, optical card technology has not fulfilled its early promise. It would be difficult to prove the confidentiality of the data or to police access to it, particularly if these systems became more common. Data on an optical card may be stored in an encrypted form, but in that case the responsibility for key storage and comparison would rest with the system, thus removing the advantage of complete portability from the card.

Current generation readers rely on software to control access to the data on the card; this is not really adequate for full-scale live applications on such sensitive data.

In the meantime, the memory capacity of smart cards has grown and the advantages of working with standardized systems and open-network technologies has outweighed the larger capacity of the optical card.

Smart cards

Origins and development

In 1970 a Japanese inventor, Kunitaka Arimura, filed the first patent for what we would now call a smart card. His patent was restricted to Japan and to the technical aspects of the invention. Japanese cards are manufactured under an Arimura license.

Between 1974 and 1976, Roland Moréno in France successfully patented several functional aspects of the smart card and sold licenses to Bull and other manufacturers. Bull further developed the microprocessor aspects of the smart card and filed several patents in its turn on this aspect of the technology. Moréno's company, Innovatron, pursued an aggressive policy of licensing and litigation worldwide, which restricted the number of companies working in this field. The main patents expired in 1996, and this has led to a rapid increase in the number of companies working in this field.

The most important aspect of the smart card, as Moréno recognized, was the control of access to the card's memory through the use of passwords and other internal mechanisms. From a technology point of view, it is important that parts of the memory are only accessible after certain specific operations have been performed. This makes the chip logic more complex, but dramatically simplifies the work of the surrounding system elements, particularly in the areas of encryption and key management.

Many chip cards, as we saw in Chapter 1, do not contain a microprocessor, but only memory governed by some fixed logic. The volume market, and the commercial success of chip card manufacturers, depended initially on telephone cards and other simpler cards.

The GSM and satellite television markets have changed the picture considerably: Although the number of memory cards sold still exceeds that of microprocessor cards, the financial success of the business is now built on the higher value microprocessor cards, performing authentication functions.

The market for all card types grew very slowly to begin with; by 1989, volumes were still less than 50 million cards a year. During the 1990s, however, growth accelerated sharply, and by 1995 the volume of memory cards sold exceeded 300 million, with microprocessor cards less than a tenth of that figure (see Figure 4.5). By 1999, total smart card shipments were around 1.5 billion a year; although memory cards still represented two-thirds of the volume, by value the microprocessor card took three-quarters of the market.

Growth in memory cards is now almost completely organic, whereas new applications are coming onstream for microprocessor cards on a regular basis. Additionally, one or two key applications, such as the high-volume transit schemes, could start moving to microprocessor

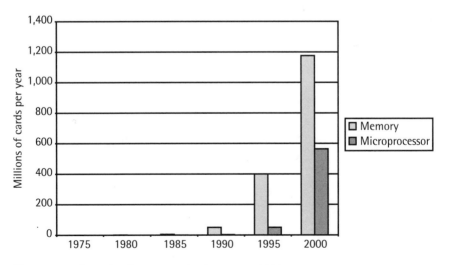

Figure 4.5 Growth of smart card sales, 1975–2000.

cards. Despite this success, these figures remain small in comparison with 10 billion magnetic-stripe cards manufactured in the world each year.

Elements of the technology

As we will see in Chapter 7, smart cards are usually the same shape and size as the other types of card mentioned earlier, although there are also other possibilities. The chip is contained within the thickness of the card, so one side is usually free for printing graphics, including microprinting and any of the other visual security features. But the security of the card resides in the chip itself, its manufacturing methods, and the data contained within it.

The card consists of a plastic carrier, in which is embedded a specially designed integrated circuit and either a set of contacts or an aerial for contactless operation. Other important elements of the system are the readers and computer systems that will use the card in operation, as well as the systems for manufacturing, issuing, and controlling the card itself and the keys that it contains. These elements are all described in more detail in Chapters 8 to 11.

Standards

A glance at Appendix A shows the profusion of standards applicable, in one way or another, to smart cards. Each of these standards will be mentioned in the more detailed sections later in the book. However, it is important to mention the several families of smart card standards that have developed independently.

The most important of these is:

- *ISO 7816:* This set of standards is a logical development from ISO 7810–7813, which covers most current magnetic-stripe cards used in banking and financial applications. ISO 7816 defines a contact card containing a microprocessor that can be used as a direct replacement for a bank card. It is a relatively low-level specification and does little to define the functions of the card.

Below ISO 7816, each of the main application sectors has developed standards or specifications relevant to its requirements. For example:

- *EMV:* Between 1993 and 1996 the major credit card schemes (Europay, Mastercard, and Visa) developed a further set of specifications based on ISO 7816, but covering the core functions of a bank card in much more detail.

- *ETSI:* The European Telecommunications Standards Institute has been responsible for a set of standards that covers smart cards for use in public and cellular telephone systems.

- *PC/SC and OCF:* These specifications have been developed to enable PC programmers to access data on smart cards, and to manipulate smart card functions, through device handlers within PC operating systems.

- *IATA:* The airline industry association has developed a specification for an electronic ticket.

Each of these standards will be described in the relevant chapters within Part 3. Although most contact-based chip cards meet the basic ISO 7816 standards at the electrical and protocol levels, application standardization has generally been much weaker. Even the ISO 14443 standard for contactless cards still exists in several flavors, roughly corresponding with the number of manufacturers, and several manufacturers still use their own memory card standards rather than ISO 7816-10.

This situation is now starting to improve. Most smart bank cards currently in issue were designed before the EMV standards were published, but new cards will conform to EMV, and in some cases also to the Common Electronic Purse Specifications, which will be described in Chapter 14. The move towards third-generation mobile telephony will increase standardization in that area, while public-transport operators are developing standards for interoperability of transit cards.

Confusion about smart card standards, and the absence of standards in many areas, contributed to manufacturers' reluctance to design terminal equipment and software for general use. This fear is probably no longer justifiable, but will persist in many areas until critical mass has been achieved by a single standard.

Hybrids

One card can make use of several technologies. In particular, many smart cards will also carry a magnetic stripe; this applies particularly to bank cards designed for international use.

When a hybrid card is used, there must be rules governing the priority sequence for trying the different technologies: Usually the chip must be tried first if the reader is able to do so. If there is no chip, or if the chip is unserviceable, then the reader may be allowed to use the magnetic stripe. The application must be aware, though, that the magnetic stripe does not incorporate the security mechanisms that are built into the chip, so extra precautions must be taken to prevent fraudsters from destroying the chip (by applying a severe overvoltage, or simply by smashing it with a hammer) and gaining access to the account by using the less secure technology.

PCMCIA cards

The growth of the market for laptop computers brought with it the need for interface cards in a very small and standardized size, distinct from the PC bus. This demand was echoed by the makers of scientific instruments and telemetry equipment, who needed not only interfaces but also portable memory.

The Personal Computer Memory Card Industry Association (PCMCIA) has produced a set of three standards, which allows not only small memory cards but also devices such as disk drives to be incorporated into a card format: slightly longer than a credit card, and up to 10.5 mm thick in the case of disk drives. PCMCIA memory cards can contain any type of semiconductor memory (it is usually one single type rather than a mixture), and many of them include a battery to sustain the memory. Memory sizes using this technology can be many megabytes.

There is very little competition between PCMCIA cards and smart cards: They each have their applications and there is little crossover between them. PCMCIA memory cards are used for transferring data and programs between laptops or instruments and a desktop computer system. No PCMCIA card, as far as we are aware, has the controlled access to memory that is the special feature of the smart card, although it would be completely practical to make a card in a PCMCIA format with a smart card chip inside it. Users of PCMCIA cards must normally make use of one of the standard PC file encryption packages and enhanced password systems to protect their data.

Others

Other technologies are used in specific sectors or applications (e.g., radio isotopes) and proprietary units such as the button memory chip. Two more important technologies must be mentioned, however; they are both commonly used for access control and other automatic identification applications, and each has some relevance to the smart card market.

Bar Coding

Bar codes are very widespread, particularly in the form of the article codes on consumer goods (UPC in North America, EAN in Europe and elsewhere) and in industrial control applications.

Bar codes are highly standardized; they can be produced very cheaply (cards or labels can be printed on site), and there is a wide variety of reader types available. The ease with which they can be produced, however, makes them inherently unsuitable for security applications: a card can be photocopied or labels printed on almost any computer printer. Infrared wavelengths can be used to give some elementary protection, but even then the codes can be reproduced using many common word-processing or printer-control packages.

In the end, however, the limitation of bar code technology is that it is read-only: It is only capable of delivering an identification number. The authentication of the medium and of the cardholder must be carried out by another method.

Bar codes are often used as a supplementary control during manufacturing and card issuing, as they can be read before the card is personalized.

Radio frequency identification (RFID)

RFID is used in access control, traffic control, and some specialized industrial-control applications. The tag, which can have many shapes, including that of a credit card, contains an antenna. When the antenna comes within range of a reader, it generates enough energy to power up a circuit in the tag and to transmit the identification number contained in the tag.

Although RFID tags are relatively difficult to copy, the limitation of the technology is again that it is read-only. There are variants of the RFID tag that allow writing, but these are usually very large and expensive and demand high power. The only version that overcomes these problems is the contactless smart card, which will be covered in Chapter 7.

5

Encryption

Cryptology overview and terminology

Cryptology is the science of codes and ciphers. It is an ancient art and has been used for many centuries to protect messages sent between military commanders, spies, lovers, and others who wished to keep their messages secret.

When we are dealing with data security, we need to prove the identity of the person sending or receiving the message and to show that the message contents have not been altered. These three requirements, for confidentiality, authentication, and integrity, are at the heart of modern data communications security, and they can all be addressed by some form of cryptology.

Cryptology has its own jargon and acronyms. A fuller understanding also requires some knowledge of mathematics. Readers who would prefer to avoid the explanation but still understand the main issues should skip to the summary at the end of this chapter.

To protect our original data (known as the *plaintext*), we encrypt them using a key so that the data cannot be understood by anyone reading it. The encrypted data (known as the *ciphertext*) appear to be a meaningless series of bits with no clear relationship to the original. To restore the plaintext, the receiving party decrypts it. A third party (such as an attacker) can use *cryptanalysis* to try to restore the plaintext without knowing the key. It is important to remember the existence of this third party!

Encryption has two main components, an *algorithm* and a *key*. The algorithm is a mathematical transformation or formula. There are few strong algorithms, and most of them are published as standards or in mathematical papers. The key is a string of binary digits, which has of itself no meaning. Modern cryptology assumes that the algorithm is known or can be discovered. It is the key that is kept secret and that changes with every implementation. Decryption may use the same or a different pair of algorithm and key.

The plaintext is often rearranged before being encrypted; this is generally known as *scrambling*. More specifically, *hash functions* reduce a block of data (which can in principle be of any size) to a predetermined length. The original data cannot be reconstructed from the hashed value. Hash functions are often required as part of an authentication system; a *digest* of the message (including the most important elements such as the message number, date and time, and main data fields) is constructed and hashed prior to encryption.

A message authentication check (MAC) is a fixed algorithm for generating a signature on a message using a symmetric key. Its purpose is to demonstrate that the message has not been altered between origination and reception. When public key encryption is used to authenticate the originating party, the result is usually known as a *digital signature*.

Algorithms

The design of cryptographic algorithms is a matter for mathematical specialists. The smart card system designer must, however, be aware of the strengths and weaknesses of the specific algorithms available and be able to decide when each is appropriate. Although cryptology has advanced immeasurably since the pioneering work of Shannon in the late 1940s and early 1950s, cryptanalysis goes hand in hand with encryption, and few algorithms stand the test of time. So the number of algorithms used in practical computer systems, and particularly in smart card systems, is very small. The main ones are described below.

Symmetric key systems

A symmetric algorithm is one that uses the same key for encryption and decryption. The most widely used form of encryption in smart cards (and indeed in almost all data security systems) is the data encryption algorithm (DEA), more commonly known as DES. DES was originally a U.S. government product, but it is now a widely used international standard (see Appendix A). Sixty-four-bit blocks of data are encrypted and decrypted using a single key, typically 56 bits in length. DES is computationally quite simple and can readily be performed using slow processors (including those in smart cards).

The scheme depends on the secrecy of the key. It is therefore only suitable in two situations: where the keys can be distributed and stored in a dependable and secure way, or where the keys are exchanged between two systems that have already authenticated each other and their life does not extend beyond the session or transaction. DES encryption is most commonly used to protect data from eavesdropping during transmission.

Forty-bit DES keys can now be cracked in a few hours using standard PCs, and so should not be used for protecting critical or long-lived data. Fifty-six-bit keys are regularly cracked by specialized hardware or networks. *Triple-DES* encryption is the process of enciphering the original data using the DES algorithm performed three times (twice forwards with the same key and once backwards with a different key—see Figure 5.1). This has the effect of doubling the effective length of the key; as we will see later, this is a critical factor in the strength of encryption.

Various newer standard symmetric algorithms have been proposed. Algorithms such as Blowfish [1] and IDEA [2] have been in use for some time, but none of these is generally available in hardware and they have therefore not challenged DES as the algorithm of choice for microcontroller-based applications. The U.S. government's Advanced Encryption Standard (AES) project has selected Rijndael to replace DES as the primary government encryption algorithm. The Twofish algorithm was specifically designed to allow implementation in low-power processors such as smart cards.

In 1998, the U.S. Defense Department made the decision to declassify the Skipjack and Key Exchange Algorithms used in its Fortezza PC card; one of the reasons given for this was to encourage more smart card implementations of these algorithms.

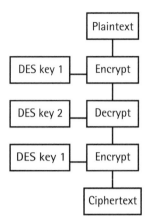

Figure 5.1   Triple DES encryption.

For *streaming encryption* (encryption of a real-time string of data in a communications network, rather than a static file), the *RC4* algorithm provides high speed and a range of key lengths from 40 to 256 bits. RC4, which is owned by RSA Data Security, is the algorithm normally used for encrypting secure sessions on the Internet (see Chapter 13).

Asymmetric key systems

Asymmetric key systems use a different key for encryption and decryption. Many systems allow one component (the *public* key) to be published, while the other (the *secret key* or *private key*) is retained by its owner. The most common asymmetric system is known as RSA (after the initials of its originators Rivest, Shamir, and Adleman), although there are several other schemes. RSA involves two transformations, both of which require modular exponentiation with very long exponents:

- *Signature* encrypts the plaintext using the secret key;

- *Decryption* performs the equivalent operation on the ciphertext but using the public key. For signature verification, we test whether this result is the same as the original data; if it is, then the signature was produced using the corresponding secret key.

The basis of the RSA system is the formula

$$X = Y^k (\mathrm{mod}\ r)$$

where X is the ciphertext, Y is the plaintext, k is the secret key, and r is the product of two carefully selected large prime numbers. For a more detailed explanation of the mathematics, see the original paper [3] or any cryptography textbook. This form of arithmetic is very slow on byte-oriented processors, particularly the 8-bit processors used in many smart cards. So although RSA will permit both authentication and encryption, it is primarily used in smart card systems to *authenticate* the originator of a message, to prove that data have not been altered since the signature was generated, and for encrypting further keys.

RSA is the basis for the authentication and key exchange functions of Pretty Good Privacy (PGP) and Riordan's Internet Privacy Enhanced Mail (RIPEM), which are systems for protecting the confidentiality of electronic mail. Both of these use symmetric algorithms, however, for encrypting the body of the text.

Other asymmetric key systems include discrete logarithm systems, such as Diffie-Hellman [4], ElGamal [5], and other polynomial and elliptic curve schemes. Many of these schemes are one-way functions, which allow authentication and verification but not decryption. A newer contender is the RPK algorithm [6], which uses a "mixture generator" to set up a combination of keys with the required characteristics. RPK is a two-phase process: after the initialization phase, it is exceptionally efficient (for a public-key scheme) at encrypting and decrypting streams of data, and can readily be implemented in custom hardware. It is therefore well-adapted to communications encryption and authentication.

Key lengths for these alternative schemes are much shorter than for RSA, which also makes them more suitable for chip card use. But RSA remains the benchmark against which other schemes are assessed; its 25 years of existence give some warranty against a major weakness.

Authentication

Authentication uses encryption to identify one or more of the parties in a transaction. Hengist and Horsa use symmetric encryption to exchange messages during the battle, having agreed on the key the previous evening. When Horsa receives a message from Hengist, with the special code attached, he is confident that it has come from his brother. (See Figure 5.2.)

This form of authentication breaks down if others can discover the key, or if we cannot exchange keys in confidence. If Romeo wants to send a message to Juliet, he can look up her public key in a directory and encrypt the

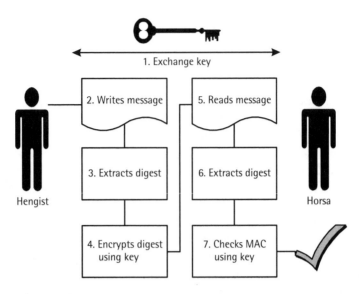

Figure 5.2 Hengist and Horsa authenticate each other using symmetric keys.

message using that key. He is then confident that only she can decrypt the message. He also takes a digest of the message and encrypts that with his private key. Juliet can look up his public key and check that the message really did come from him. (See Figure 5.3.)

A particularly elegant and efficient scheme for authenticating a smart card within a system is the zero-knowledge scheme proposed by Guillou and Quisquater [7]. This involves a two-part challenge-response mechanism, as shown in Figure 5.4.

In a zero-knowledge test, the entity being authenticated is able to prove its identity (its knowledge of a secret) without giving away the secret. The shadow ID used in the Guillou-Quisquater scheme is much longer than the card ID, while the secret number is a modular function of the card ID. Again, the computation requires modular exponentiation, but exponent sizes can be kept relatively short because the challenge is only used once and there is no plaintext for comparison.

Keys

Having established the importance of keys to data security, we must now consider what different types of keys can be used and where they are appropriate.

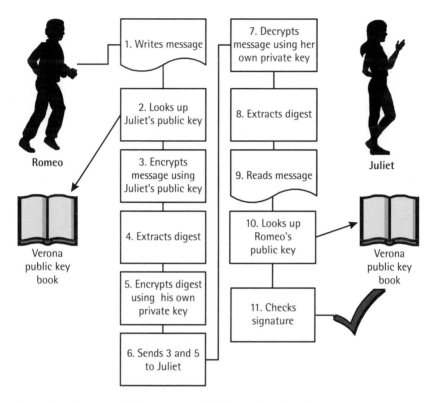

Figure 5.3 Romeo and Juliet use public key authentication.

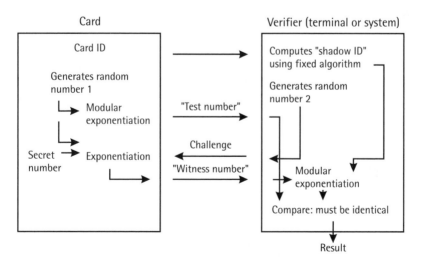

Figure 5.4 Guillou-Quisquater zero-knowledge test.

Secret keys

Symmetric algorithms such as DES use secret keys; the key itself must be transmitted and stored by both parties to the transaction. Because we assume that the algorithm is known, this places great emphasis on the need to transmit and store keys securely. Smart cards are often used to store secret keys. In this case it is important to ensure that the scope of the key is limited: We should always assume that one card may be successfully analyzed, and this must not put the whole system in jeopardy.

Public and private keys

The main advantage of asymmetric key systems is that they allow one key (the private key) to be held very securely by its originator while the other can be published. Public keys can be sent with messages, listed in directories (there is provision for public keys in the ITU X.500 electronic messaging directory scheme), and passed from one person to the next. The mechanism for distributing public keys can be formal (a key distribution center) or informal (the term *web of trust* is used by some authors).

The secrecy of the private key in such a system is paramount; it must be protected by both physical and logical means within the computer in which it is stored. Private keys should never be stored unencrypted in a conventional computer system, nor in human-readable form. Again, smart cards are used for storing private keys for individuals, but private keys for major organizations should not normally be stored in a card.

Master keys and derived keys

One way of reducing the number of keys that must be transmitted and stored is to derive the keys each time they are used. Under a derivation key scheme, a single *master key* is used, together with some individual parameters, to calculate a specific derived key, which is then used for encryption. For example, if a single issuer has a large card population, the card number may be encrypted using the master key to produce the derived key for each card (see Figure 5.5).

Another form of derived key is calculated using tokens, which are electronic calculators with special functions. They may use as inputs a challenge issued by the central system, a PIN entered by the user, and the date and time. The token itself contains the algorithm and a master key. Such tokens are often used for secure computer system access. An example token is shown in Figure 5.6.

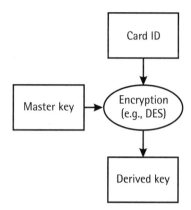

Figure 5.5 Key derivation using a master key.

Figure 5.6 Secure access token. (Photograph courtesy of RSA Security, Inc.)

User and equipment keys

This principle may be extended to produce complete hierarchies of keys. For example, a function (setting up staff IDs or an order file) may only be allowed by one group of users from one specific terminal or group of terminals. Critical parts of this function (e.g., authorizing the production of a staff card or transmitting the orders) may require one individual's personal authority. In this case, we might use a hierarchy such as that shown in Figure 5.7.

Key-encrypting keys

Because the transmission of the key is potentially a weak point in the system, it makes sense to encrypt keys when they are transmitted and to store them in encrypted form. The key-encrypting keys never pass outside the computer system (or card) and can therefore be protected more easily than the transmission itself. Often a different algorithm is used to exchange keys from that used to encrypt messages.

We use the concept of key "domains" to limit the scope of keys and to protect keys within their domain. Typically a domain will be a computer system that can be protected logically and physically. Keys used within one domain are stored under a local key-encrypting key for that domain; when they are transmitted to another computer system, they are decrypted and reencrypted under a new key, often known as the *zone control key*. On receipt at the other end, they are again translated to encryption under the new system's local key (see Figure 5.8).

Session keys

To limit the time for which keys are valid, a new key is often generated for each session or each transaction. This may be a random number, generated by the terminal following authentication of the card (see Figure 5.9). If the card can perform public-key decryption, then the session key can be encrypted using the card's public key.

The part of the transaction in which the key is transmitted is often quite short in comparison with the rest of the transaction; thus, it can carry

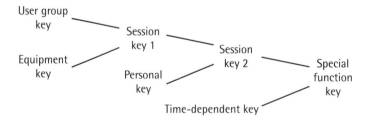

Figure 5.7 Example of a hierarchical key structure.

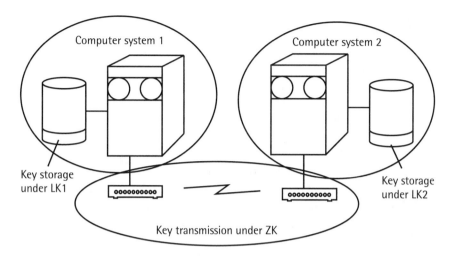

Figure 5.8 Key domains and key encryption.

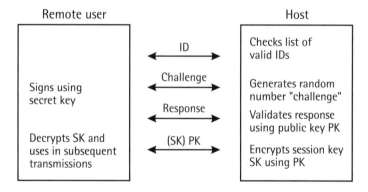

Figure 5.9 Authentication and key exchange dialog using asymmetric keys.

a higher overhead than the bulk of the transmission. The processing time for the authentication and key transmission phase is acceptable if the rest of the transaction can be encrypted with the lower overheads of the symmetric key.

A special form of session key is the transaction-key system, used in some electronic payment and electronic data interchange (EDI) systems. Here, a new key is transmitted at the end of each transaction, and this is used for the next transaction.

Selecting an algorithm and key length

No cryptosystem is absolutely secure under all circumstances. There is always a chance that someone may stumble on the algorithm and the key by pure chance. Or, particularly where the algorithm is known, an attacker can try every key combination until one appears to work (a *brute force* attack). This task becomes more time consuming as the key becomes longer, and it should be possible to calculate the time required to do an exhaustive search of the whole keyspace (i.e., all possible key values). For a strong algorithm and suitable key length, this may be several decades of processing using the most powerful supercomputers currently available (although the right answer could be the first key tried!).

The strength of any particular form of encryption depends on the length of the key. This is not only because the brute force attack takes much longer, but also because there is less scope for speculative attacks: feeding in possible combinations of values or trying likely plaintext values.

The criterion for selecting an algorithm and key length is that the cost and effort required to find the key or break the security should be greater than the maximum possible reward.

This criterion is important for card-based systems. With a limited range of algorithms and key lengths available, the other variable within the system designer's control is the maximum reward. In other words, the system must limit the damage that could be done by breaking a single code, so that it is not worth an attacker's time trying to break it. The main ways in which this is done are to:

◆ Limit the scope of a single key (to one card or one type of transaction).

◆ Limit the time for which a key is valid. Many keys are only used for a single message or transaction; it is easier to protect a key that is only used to generate other keys than a key used for encryption.

◆ Include sensitive monitoring functions in the system, so that any irregularity is quickly detected and the cards or parts of the system affected can be isolated.

When symmetric encryption is being used, therefore, it is best to generate keys in a hierarchy, so that only the lowest levels are liable to be compromised. The DES standard (and hence all generally available hardware) uses 56-bit keys. Versions used outside the United States often use "weak" forms of DES, with 40-bit keys, although full-strength versions developed outside the United States are available. Forty-bit DES can now be broken on single PCs in little more than a day, and even 56-bit DES has now been broken in less than a day using a network of 100,000 PCs on the Internet [8]. Where a DES key is used repeatedly, therefore, the triple-DES technique (using two keys) or a different algorithm (such as Blowfish) should be used. The alternative is to generate a new key for each message or transaction. Banking message encryption has used triple-DES for many years.

With public-key systems, the length of key required depends entirely on how the key will be used. In general, the scope for discovering the private key is usually much more limited than for a symmetric key, because it should never be used directly except within the confines of the key owner's own computer. Only I may use my key. However, the private key could be open to a brute force or analytical attack if I have used that key to authenticate (by signing) many messages or other data, and the attacker gains access to examples of the signed data and the original.

Public-key systems based on the RSA principle depend on a number that is the product of two large prime numbers. If this number can be factored into its components, the key can readily be broken. So the key should be as long as practical, subject to the maximum acceptable transaction time in the application concerned. For access to a secure building or computer, for example, a transaction time of several seconds might be acceptable, but when boarding a bus anything over 200 ms results in a large number of failed transactions and impatient passengers.

For comparison, an RSA key of 64 bits was broken in 1995 using three linked, massively parallel computers over four days; a 40-bit key in 1996 required 3.5 hours with a network of 250 workstations. The time required for factorization increases exponentially with the length of the key (a 65-bit key should take twice as long as a 64-bit key), so unless new factorization methods are discovered (this is an important reservation) the keys used today, typically 768 or 1,024 bits, still have an adequate margin of safety.

The way the RSA scheme is operated, however, also affects its security: The French banks decided in 1983 on a simplified scheme that involved a 320-bit "secret number." This number was broken by factorization in 1999, and was used to create counterfeit cards that could respond correctly to the authentication challenge used by the off-line terminals widely used in France. This factorization was predicted as long ago as 1988 [9], although at that time the longest number which had been factorized was 299 bits (in several days on a Cray supercomputer). But the cards remained many times more secure than the magnetic stripe cards they replaced, and more powerful cards would not only have been more expensive, but also slower.

Whatever the scheme used, it remains important to ensure that the key is not compromised by other means. Today's bank card schemes are much less vulnerable than the 1983 French scheme, because the bank and card-scheme keys have minimal exposure. This is the purpose of key management.

Key management

As with the locks on a building, users of encryption systems must make sure that their keys are created and stored securely, and that they are only available to properly authorized personnel.

Key generation

Symmetric keys should normally be generated completely randomly. Many so-called random-number generators generate not a random number, but a fixed sequence. This characteristic can be useful in some situations (such as the *code guns* used in radio systems), but if the source and starting point can be determined, then the entire system is compromised. True random-number generators typically digitize some of the *white noise* generated by a free-input amplifier.

Asymmetric key generation is a more complex process, as it requires the use of large prime numbers and not all combinations are suitable. The process must still be seeded using a random input, however, or the same problems will ensue. In addition, it has been found that some combinations of keys are easily broken, and the key set produced must be tested for these conditions before it is used. These processes are normally carried out automatically within the key-generation system that forms part of every public key encryption package.

Some chip cards with advanced security functions are now able to generate a public-private key pair on the card. The private key never leaves the card. This is more secure than the conventional method, in which the private key is loaded into the card during prepersonalization (see Chapter 11).

The process of key generation must be carefully controlled. In many organizations, it takes the form of a small ceremony: Several people are involved and each checks the operations of one other. No one person has access to the whole sequence.

Key transmission

The problems of key transmission have already been mentioned and some of the solutions described. Regular transmission of session keys is handled by encrypting them under a key-encrypting key during transmission. Where a master key must be loaded into several pieces of equipment, it is now normal to connect them to a key server and to load the keys directly, using the same technique.

In some systems, it is still necessary to transfer keys manually. In this case it is usual to divide the key into three or more parts; each part is transferred to paper (or committed to memory) by one person and loaded by the same person at the other end. Whether the process is carried out manually or automatically, a high level of formal control and discipline is required during this process.

Key transmission is less of a problem with asymmetric key systems because only the public key is normally transmitted from one system to another. Even then, it is desirable to use a MAC or some other form of integrity check to ensure that the key is not altered during transmission.

Key indexes

By now it will be clear that there is an underlying assumption that keys may be discovered, even if we do everything we can to avoid this. We must have a plan for detecting compromise and know how we will react if this happens. We have also said that one way to limit the damage caused by a breach of security is to limit the time for which any key is valid. Both of these requirements point to the need to set up more than one set of keys and to have a mechanism for moving from one set to the next.

This is done by generating several sets of keys, each of which is associated with an index. When the key is stored, the index remains with it. Cards and other remote systems may be loaded with three or more keys, and they advance to the next key when instructed to do so by the host system (for off-line systems such as smart cards, this will take place at the beginning of the next transaction). The new key set need never be transmitted.

There is a particularly high danger of key compromise during the testing phase of any system, so it is wise to advance the index after tests are completed. And when any major improvements to security have been made, the benefit of the improvement may be lost if the existing keys remain in use, so this is another occasion on which a new set of keys should be used.

Certification authority

A *certification authority* (CA) is a trusted computer system that is able to testify to the identity and authenticity of one or more of the parties in a transaction. Often this is the owner of the scheme. Certification can be an on-line or an off-line process; either the entity presents its ID together with a certificate from the CA or it presents its ID and the other party seeks certification on-line from the CA. The process may be mutual: Each party may require authentication of the other.

If two parties do not know each other in advance, they may use a CA as a trusted third party to confirm each other's identity. If Juliet does not know Romeo's public key, she may ask Friar Lawrence to confirm its authenticity, or to verify her signature to Romeo (Figure 5.10).

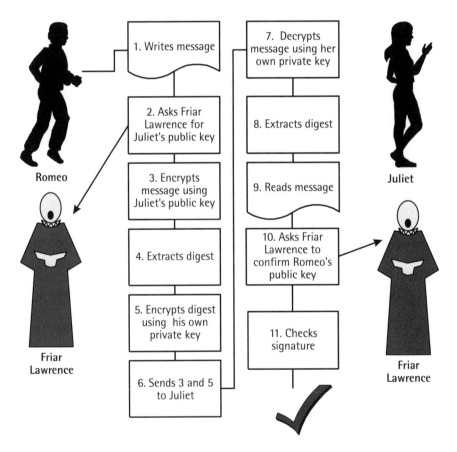

Romeo

1. Writes message

2. Asks Friar Lawrence for Juliet's public key

3. Encrypts message using Juliet's public key

4. Extracts digest

5. Encrypts digest using his own private key

Friar Lawrence

6. Sends 3 and 5 to Juliet

7. Decrypts message using her own private key

8. Extracts digest

9. Reads message

10. Asks Friar Lawrence to confirm Romeo's public key

11. Checks signature

Juliet

Friar Lawrence

Figure 5.10 Using Friar Lawrence as a trusted third party.

Where authentication is off-line, there should usually be a mechanism for registering invalid certificates (where an organization no longer belongs to the certified group). This may be done by giving certificates a limited life or by transmitting lists of lapsed certificates to those still subscribing.

The CA may also act as a central repository or distribution center for public keys. To guard against incorrect public keys being published or errors being introduced, public keys may be signed by the CA, using its secret key. They can then be decrypted using its public key, which is assumed to be available to all.

Public key infrastructures

Where entities (people, cards, or computers) need to authenticate themselves outside their own circle or closed user group, we use a hierarchy of CAs, known as a public key infrastructure (PKI).

Every bank cannot know every other bank, still less every other bank's customers. So customers are certified by their own bank, banks are certified by a national interbank organization, and for international transactions interbank organizations are certified by an international clearing house or card scheme (see Chapter 14).

Companies trading with each other use PKIs set up by Chambers of Commerce or by a trading exchange they each belong to. Their employees, who have individual private keys, have their public keys certified by the companies. The employees and computer systems are able to trust each other by checking the sequence of certificates up and down the chain.

The existence of such public key infrastructures is one of the key requirements for automating business processes. However, setting up widespread PKIs raises many issues. For example:

* Who sets up and controls lateral links, for example, between banks and telecommunication companies for payment over mobile telephones?

* Government involvement: Many governments believe that only they should issue what are in effect certificates of identity to individuals. Certainly governments are unlikely to accept certificates issued by commercial organizations for use in public life.

* How are certificates revoked and upgraded? A scheme might work well for several years but fall apart when large numbers of certificates start to expire.

- How are security breaches detected and what action can be taken?

- What liability is taken on by a CA? What responsibility does it have towards a person relying on a certificate?

These questions are normally addressed by a CA in its certification practice statement (CPS), which forms an essential part of its trading conditions. But the answers seek in most cases to avoid responsibility rather than resolve the issues. Users of chip cards are rarely aware of the CPS that underpins the certificates on their cards.

Computational requirements

All cryptology requires mathematical manipulation of blocks of data. Even general-purpose computers can be slow at these computationally intensive tasks, and various functions can help in this process.

DES is available directly in hardware: With its fixed length input and key size, it is well suited to large-scale integration, either on its own or in conjunction with other cryptographic functions. Several manufacturers offer this function.

For RSA and similar algorithms, there are two possible approaches: to provide a fast arithmetic unit (which provides rapid execution of multiple-length arithmetic functions), or to provide special functions adapted to the solution of expected encryption problems.

Most specialist cryptographic hardware (including cryptographic micro-processors used in smart cards) now adopt the second approach. Although it is less flexible, the performance improvement over the general-purpose coprocessor is very substantial. Cryptoprocessors are able to incorporate special functions such as the Chinese Remainder theorem, which provides a fast route to the modular exponentiation required by RSA.

The performance of smart cards when performing cryptographic functions will be discussed in more detail in Chapter 8, but chips are now available (for use in terminals and host systems) that will perform triple-DES calculations in under 100 ms and 1,024-bit RSA in under 200 ms.

DES is also widely available in software form, including several free-ware sources. Although several of the DES standards exclude software implementations in general-purpose computers, these are nonetheless widely used. The alternative Blowfish algorithm is freely available and downloadable from www.counterpane.com/blowfish.html.

Cryptography export controls

Many governments treat encryption products as weapons and seek to control their use and particularly their export. The United States and French governments have both relaxed their stance in recent years, but the United Kingdom retains particularly stringent regulations. The data communications industry has fought against these restrictions, arguing that their effect is to give criminals access to stronger security than law-abiding citizens.

The wide availability of many algorithms in published—and even in software—form means that the controls are not completely effective, although they do constrain the international flow of some implementations and restrict the length of keys that may be used in other cases. Most commercially available products use licensed forms of the relevant standards and are not subject to further controls.

From a practical point of view, much of the most advanced work on cryptography is carried out in government laboratories such as the National Security Agency (NSA) in the United States and the Computer and Electronics Security Group (CESG) in the United Kingdom. They are able to control what is published.

The U.S. government sought in the early 1990s to impose a hardware-based encryption scheme known as Clipper. This was fiercely resisted by the cryptology community, and Clipper is now only mandated in a small number of cases.

Other governments have sought to impose key-escrow arrangements, under which any user of "strong" encryption products is required to lodge all keys with the appropriate government departments. This has so far proved impractical to implement, and it raises a wide range of opposition from civil liberties and consumer protection groups.

Summary

For the smart card system developer, a few key facts can be extracted:

- ◆ There is an encryption method available to handle almost any security requirement. The normal systems designer should make use of commercially available products rather than trying to design a unique system; such products will be licensed and meet the relevant requirements for encryption usage and export.

- The requirements for card authentication, data confidentiality, and message integrity should normally be considered separately. Authentication can be performed using a symmetric algorithm such as DES or an asymmetric (public-key) system such as RSA. Data stored within a suitable chip card can be considered confidential provided that it is protected by an appropriate access-control mechanism. However, if it passes outside the card, it may be encrypted, probably using DES or another symmetric algorithm. Message integrity should be ensured by using a standard message authentication check.

- DES is computationally simpler than RSA; it is available in hardware and in several smart card chips. However, it requires a secret key to be shared between the card and the decrypting device; thus, it is not generally suitable for card authentication in open systems.

- RSA requires more computation, and is only available at reasonable speed in the fastest smart card chips; however, the ability to publish one key openly, while the other key remains secret inside the card or host system, is the reason it is more commonly used in open systems. Simpler chips can store a certificate created with the secret key, which can be checked by a terminal.

- The strength of any encryption system depends on the length of the keys used; the keys used in typical applications will be discussed further in Part III.

References

[1]   Schneier, B., "Description of a New Variable-Length Key, 64-Bit Block Cipher (Blowfish)," *Fast Software Encryption, Cambridge Security Workshop Proceedings,* Cambridge, United Kingdom, December 1993, New York: Springer-Verlag, 1994, pp. 191–204.

[2]   Lai, X., and J. L. Massey, "A Proposal for a New Block Encryption Standard," *Advances in Cryptology, Proceedings of Eurocrypt 1990,* Springer-Verlag, 1990.

[3]   Rivest, R., A. Shamir, and L. Adleman, "A Method for Obtaining Digital Signatures and Public Key Cryptosystems," *Communications of the ACM,* Vol. 21, No. 2, 1978.

[4]   Diffie, W., and M. E. Hellman, "New Directions in Cryptography," *IEEE Transactions on Information Theory,* Vol. 22, No. 6, 1976.

[5]   ElGamal, T., "A Public Key Cryptosystem and a Signature Scheme Based on Discrete Logarithms," *IEEE Transactions on Information Theory*, Vol. 31, No. 4, 1985.

[6]   Raike, W., "The RPK Public-Key Cryptographic System," published on http://crypto.swdev.co.nz.

[7]   Guillou, L., and J. -J. Quisquater, "Zero Knowledge Identification Proof for Smart Cards," *EUROCRYPT '88, Workshop on the Theory and Application of Cryptographic Techniques*, Davos, Switzerland, May 1988, *Lecture Notes in Computer Science*, Springer-Verlag.

[8]   "DES Challenge III Broken in Record 22 Hours," press release, Electronic Frontier Foundation, January 19, 1999.

[9]   Guillou, L., *Annales de Télécommunication*, No. 43, 1988.

6

Passwords and Biometrics

This chapter is concerned with the many situations in which we need to be able to identify, or verify the identity of, a person. We are mostly concerned with checks that can be carried out automatically by a system, although the system may also be used to provide the evidence for another person to check.

Personal identification types

Passwords, tokens, and biometrics

Virtually all identification methods involve something you *know* (a password or PIN number), something you *have* (a card or other token), or something you *are* (a physical or behavioral characteristic). The last group is known as biometric checking.

Each has its advantages and drawbacks. Passwords can be learned by subterfuge, guessed, or given away, but they do allow the user to delegate authority. (Elderly or disabled people often ask others to draw money using their ATM cards, and it is sometimes useful to be able to ask someone else to look up some data on our computers when we are out of the office.)

Many people simply forget passwords, particularly if they use them infrequently. Tokens can be lost or stolen, or again they can be lent or transferred to someone else deliberately. Physical characteristics are inflexible; they are often difficult to transmit down telephone lines, for example. Security system designers must ask themselves whether users should have the ability to delegate or transfer their rights to someone else, as this will affect the choice of identification type.

Some identification methods can only be used in a face-to-face mode, when both the person seeking acceptance and the person checking the identity are present. Others may be used when one or the other is absent: for example, for cash withdrawals at an ATM or for remote access to a computer system. In this chapter we are mostly concerned with methods that will allow an automatic check, without any human involvement, as this is the form of identification normally addressed by smart card systems.

Identification methods can be used in combination: A card and password or card and biometric would be common. The combination may be varied according to the requirements: for example, we might use a card only for entry to the building, a card and a PIN for the computer room, but a card and a fingerprint for a funds-transfer operation on the computer system.

Where several characteristics are known, the combination can be varied dynamically: The user presenting a card might be asked for additional identification depending on the transaction or the results of previous checks. This is particularly useful in avoiding false rejects (see the next section).

Behavioral and physiometric

Biometric checks are divided into:

- *Behavioral techniques* that measure they way we do something such as signing our name or speaking a phrase;

- *Physiometrics* that measure a physical characteristic such as a fingerprint or the shape of a hand.

Behavior changes with time and mood. Behavioral measurement techniques work best when they are used regularly, so that these changes, and each individual's level of variability, can be taken into account. The models for behavioral measurement have to take into account these ranges.

Physiometric measurements, on the other hand, require larger measuring devices and more complex software. They have to match the position of the hand, for example, with the template (it will never be identical).

There are psychological differences between the two groups as well. With a behavioral method, users typically feel that they are in control. With a physiometric system, on the other hand, users sometimes feel threatened by something that denies them entry or use based on a measurement they cannot control.

Requirements

Recognition versus verification

We must first distinguish between systems designed to recognize people (is this a person I recognize and, if so, whom?) and those that must only confirm a person's identity (is this the person he or she claims to be?). This second task is much easier and the verification parameters can be set on an individual basis. This is also the normal case for smart card systems, whether the reference pattern (usually called the *template*) is held on the card or on a central system.

Performance

The most important performance characteristics of a biometric are its ability to reject impostors while accepting all valid users. These are usually measured by the false-accept rate (FAR)—the percentage of impostors accepted by the system—and the false-reject rate (FRR)—the percentage of valid users rejected. Most systems can be tuned to provide either sensitive detection (low FAR but high FRR) or coarse detection (low FRR but high FAR)—see Figure 6.1. The critical measurement is known as the *crossover rate*—this is the level at which the FAR and FRR are identical. Most commercial biometrics today have crossover rates below 0.2%, and some are below 0.1%. The rate improves with frequency of usage, as users become accustomed to the system and the system becomes better tuned to the level of variability expected.

The balance between these two extremes depends on the application: Will it be used by the public or by trained employees? Are there customer-service issues or do users have little choice? Typically, banks seeking an ATM biometric will opt for a higher false-accept rate to avoid offending valid customers, whereas a welfare agency checking applicants will demand

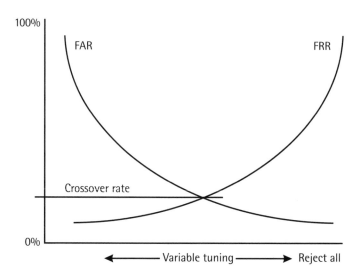

Figure 6.1 False acceptance versus false rejection in a biometric.

a very low false-accept rate, even if this means extra time to check those rejected.

Procedures

A biometric test normally involves three stages: enrollment, use, and update.

Users are enrolled in the system by an initial measurement. This will usually be performed three or more times to check for consistency; the procedure often takes longer than the measurement for testing purposes.

In normal use, the user must be guided to help provide consistent results; it is no use, for example, enrolling a signature on a large plate placed at a convenient angle and then expecting similar results from a field unit that is much smaller and located under a privacy shield. Once the measurement is captured, it is compared with the template; this is where it is important to allow appropriate levels of tolerance, particular for behavioral measurements.

Most biometric systems, particularly those that use a behavioral characteristic, must have a provision for updating the template. In the case of signature and voice recognition, this is usually an adaptive function, performed by the software every time the measurement is checked. For slower moving characteristics, the system may monitor the percentage fit or the number of times a person is rejected and seek a reenrollment when necessary. A transaction log is often a useful feature and can easily be incorporated in a smart card–based system.

Components

A biometric system will include:

- *A measurement device*, which forms the user interface. Ease of use is another important factor for biometrics: The device must be instinctive and leave little room for error. It must be suitable for use by a wide range of people, including those with disabilities.

- *Operating software*, including the mathematical algorithms that will check the measurement against the template. The newest algorithms depend less on statistical modeling, and more on dynamic programming, neural networks, and fuzzy logic. This increases their flexibility; they are less likely to reject someone because of a patch of dirt, for example, if the rest of the pattern fits closely.

- *External hardware and systems:* the usability, reliability, and cost of the system will often depend at least as much on these external systems as on the measurement device. Some systems (such as fingerprint checking) are intrinsically well suited to use in distributed systems, while others (such as voice recognition) are more appropriate for centralized systems.

Training and user instructions also form an important part of a successful biometric implementation. Where the system must be used by the public, the emphasis will be on the user instructions and design of the measurement station; with staff and other closed user groups, more benefit will be obtained from training.

The cost of biometric devices is falling rapidly. Whereas only five years ago, most biometric techniques cost around $1,000–$3,000 per station, today's fingerprint readers and iris scanners (see later in this chapter) are available for under $200. At this level, it becomes practical to fit them to ATMs and specialized access-control devices. A further drop in cost, to below $100, will probably be required before biometrics will be used regularly in a retail, vending, or mainstream access-control environment.

Passwords and PINs

The best known and most common form of personal identification is the password. Passwords on their own are not well suited to high-security applications: People forget them, write them down, or use easily guessed words, and they can be monitored and replayed. A very high proportion of successful computer system hacks involve the discovery of passwords by one means or another.

Much better results are available when passwords are combined with some form of token or challenge-response generator, so that the password is not input directly to the system. This removes one of the biggest advantages of the password—its cheapness—and does not overcome the limitations of users' memories.

A PIN is a simple form of password, consisting usually of four to six digits, which can be used with a numeric keypad rather than a full keyboard. They are familiar to most of us from their use in conjunction with bank cards. Many people cannot remember even a four-digit PIN, particularly if they have several cards with different PINs; thus, the ability to change a PIN is important. Many users write their PINs down and keep them close to the card, where they can find them when required. All of these factors reduce the value of the PIN as a safety mechanism, but they can be overcome by suitable education of users. Nevertheless, PINs should only be regarded as a secondary identity check; the card is the primary identification.

Passwords and PINs share the advantage that they are either right or wrong: There is no latitude for error and hence the software required to check them is very simple. The opposite is true for most biometric checks.

Behavioral

Signature verification

Automatic signature checks are a natural extension of a familiar process: No cultural change is required. Whereas the human operator checks the final shape of the signature, however, most automatic forms of signature verification place much more emphasis on the dynamics of the signature process: the relative speed with which lines are drawn and the pressure used. This enables them to compare measurements between signatures even where the environment is quite different and defeats most attempts at forgery.

There are many proprietary methods for measuring and storing signature dynamics, several of which have been patented. Each device measures different aspects of the signature and attempts to differentiate between stable aspects and those that vary from one signature to the next. The measurements are made using a digitizing tablet or a special writing surface.

As with all behavioral characteristics, signatures are subject to the mood of the user, the environment, the pen and paper, and so on. Some people's signatures are very consistent, where others vary greatly. By careful system tuning, very good results can now be obtained for regular users, but there are often a few "problem users" who are consistently rejected.

The signature template is typically 1 KB; this makes the technique suitable for on-line use or with a smart card. A large-scale pilot was very successful in reducing benefits-claim fraud in a U.K. unemployment office.

A subsidiary advantage of most signature-verification systems is that they capture a record of the signature as proof of the transaction: This allows *truncation* of paper systems such as credit card vouchers, since the electronic record of the signature removes the need for the paper voucher.

Keystroke dynamics

The way a person types at the keyboard is almost as characteristic as the way they sign their name; skilled typists in particular can be recognized almost instantly from their typing patterns. This offers a quick way to identify computer-system users, without requiring any additional input device or enrollment procedure. Current implementations are limited to the laboratory, due to the problems of variations between keyboards and system software delays. On the other hand, the low overhead and transparent operation make this a very attractive technique for applications such as protecting a small number of supervisor-status users on a computer system. Although the applications of such a biometric are clearly limited, this is potentially a very useful additional tool in the armory of computer access-control techniques.

Voice recognition

Voice recognition systems are the biometric most readily accepted by customers, but unfortunately they have not yet reached the levels of performance required in most commercial environments.

Using voice recognition allows multilevel checking: The system can check what is said as well as how it is said. In some environments it is very low cost: A single digital signal processor (DSP) can process for many microphones. Voice recognition is a branch of speech processing technology that has many wider applications in other fields, notably digital telephony and videoconferencing. Use is made of spectral analysis and specialized algorithms that are being developed for high-density compression and accurate representation of speech.

High levels of false rejection are a problem with voice recognition systems, often due to colds or unusual background noise conditions, but one neural network system has achieved a crossover rate of better than 0.1 %. Impersonation is less of a problem than might be thought because the aspects of the voice measured by the systems are not the same as those a human listener is likely to notice, whereas an impersonator concentrates on the human characteristics. One of the most important factors in voice recognition is the size and shape of the "voice box" in the throat.

Voice recognition can be used over telephone lines and has been used by Sprint in the United States since 1993. Although it could be used with smart cards, we are not aware of any commercial schemes that use this combination yet.

Physiometric

Finger/thumbprint

This is the oldest form of remote identification test; although it was previously associated with criminality, research in many countries is now showing a level of acceptance that would allow its use in public schemes. The recent availability of volume-produced devices for use with PCs will help acceptance.

Systems can capture the minutiae of the fingerprint (points such as intersections or ridge endings) or the whole image. Minutiae are usually captured in the form of x and y coordinates and a direction; if these three variables are used, the resolution can be quite coarse. This leads to templates of around 100 bytes, compared with 500 to 1,500 bytes for a full image of the fingerprint. A differential method has also been developed; this produces interference fringes which can be encoded in an even smaller template (around 60 bytes). Crossover rates of less than 0.05% can be achieved.

There are some problems with consistency in public schemes; workers in some trades, as well as heavy smokers, often have very faint prints which are difficult to analyze repeatably. Nonetheless, there have been many successful and long-running schemes using fingerprints, including a high-profile trial at the World Expo in Seville in 1994.

Special single-chip sensors are now available for this application; they measure variations in capacitance across an array the size of a finger, and have been built into several devices, including a smart card reader (see Figure 6.2) and a mouse (see Figure 6.3).

Fingerprint readers are now available in a wide range of packages for accessing PCs and laptops, often in conjunction with some form of card reader. Although currently priced at a level attractive only to the power user, fingerprint readers are already available in normal PC shops and volume will drive prices down quite rapidly.

Hand geometry

Hand geometry scores highly for ease of use because the hand is a large feature and can be placed very consistently using a guide system. Hand-geometry measurement devices of different sorts have been in use since the early 1970s, making this the oldest form of biometric. The hand is guided into the correct position for scanning using a set of pegs, and the image is

Figure 6.2 Fingerprint smart card reader. (Photograph courtesy of Precise Biometrics AB.)

Figure 6.3   Fingerprint mouse. (Photograph courtesy of Siemens AG.)

captured using a CCD camera. The reference template can be made very small (the most widely used commercial product uses only 9 bytes).

Although day-to-day variations such as dirt and cuts do not affect its performance, this measurement is affected by injury and aging, and if the template cannot be updated dynamically, reenrollment may be required from time to time.

High-profile applications have included access to the athletes' village at the 1996 Olympic Games in Atlanta, frequent traveler controls at several U.S. airports, and for voting in the Colombian legislature. Hand geometry is also used for many other access-control applications.

Retina scan

Retina scanners measure the characteristic blood vessel patterns on the retina (the back of the eye) using a low-power infrared laser and camera. This technique involves placing the eye close to the camera to obtain a focused image. It has been used for several years to control access to a number of U.S. military establishments. Templates are very small (35 bytes in the most common commercial system) and the crossover false-accept/false-reject rate is very low.

Recent medical research has shown, however, that retina characteristics are not as stable as was once thought: They are affected by disease, including diseases of which the owner may not be aware. Many people are worried at the thought of placing their eye in close contact with a light source and the damage this might cause. As a result, this technique has largely given way to the less invasive iris scan.

Iris scan

Iris scanners measure the flecks in the iris of the eye. The technique has a very high level of discrimination as a result of the large number of features and is found to be very stable with time.

The user has to look at a camera from a distance of 30 cm or more for a few seconds. The system accommodates spectacle and contact lens users without difficulty, although the sensor must be mounted or adjusted in such a way that it is suitable for users of different heights, including those in wheelchairs.

Iris scanning is a newer technology than most of those described; it is used in several U.S. prisons, where the exceptionally low FAR is a major advantage, and is now being tried in ATMs.

Others

Almost any part of the human anatomy could in theory be used in a biometric. Commercial schemes have concentrated on those that are easily measured and are likely to be most socially acceptable.

The story of the Indian family who collected the father's pension for years after his death by using his finger may be apocryphal. Nevertheless, it is sometimes necessary to check that the measurement is being taken from a live person, rather than a copy. Some devices include checks for this (such as movements in the flecks of the iris or subcutaneous structures). One method that uses a living characteristic directly is vein scanning: This measures the position of the veins from the flow of warm blood.

Facial recognition—the technique humans use most—is a viable biometric, particularly where cameras are already in use and where a very coarse check is all that is required. Specific features or hot spots are measured and used to create the template.

The ultimate biometric is DNA analysis. However, it can safely be said that it will be many years before this technique is used for checking identity in the average store or when boarding a bus!

Biometrics and cards

Some identification techniques, notably passwords and PINs, are exceptionally well suited to implementation in a distributed system. They have minimal storage and processing requirements, but as we have seen, they are not regarded as secure on their own.

Many systems make use of a card in conjunction with another technique. The minimum requirement is to provide a reference or identification number (this is who I am claiming to be); the template or reference password is held on a central system. However, in other applications, there are advantages in making the template portable. This allows the check to be carried out by off-line terminals, and a cardholder's ability to carry his or her own template can make the system seem less like Big Brother is in control.

The storage requirement of the template is thus a limiting factor; as we have seen, most templates require between 40 and 1,500 bytes. The smaller templates can be stored on a magnetic-stripe card or bar code, albeit with the limitations of those technologies mentioned in Chapter 4. For larger templates, two-dimensional bar codes or holograms may be used, but the most satisfactory answer in almost every case where secure storage is required is to use a smart card.

Cards are particularly useful for combining biometrics: If two or three biometrics are stored on the card, false rejects can be minimized by calling for a backup reading. Or the same card can be used in different environments: voice recognition for telephony, fingerprint at a workstation or ATM, and a PIN at the point of sale. Such "layered" biometrics are now starting to be used in commercial applications.

In the next three chapters, we will explore the characteristics of the smart card that make it particularly suitable for storage and checking of biometrics and passwords.

7

Chip Card Types and Characteristics

Chip cards fall into several categories. Within each category there is a range of chip capacities, power, and speeds. In this chapter we consider the broad product groups, and in Chapter 8 we look at the quantitative factors that distinguish chip card products. We are primarily concerned here with differences that affect security; there are several other publications, some of which are listed in the bibliography, that describe other smart card characteristics.

Chip cards can be divided into groups by:

- *Function:* The fundamental difference is that between memory cards and microprocessor cards, and there are further specialized functions within each group;

- *Access mechanism:* Contact or contactless;

- *Physical characteristics:* Shape and size.

Memory cards

Unprotected

The simplest chip cards are not smart at all: They contain memory circuits that are directly accessible through the contacts using a synchronous protocol. Some early telephone cards used this scheme, but they are not widely used today.

Protected

Most current memory cards contain, as a minimum, a protected area. The memory in the card—usually electrically erasable programmable read-only memory (EEPROM or E^{2}PROM)—is divided into two or more areas.

One part is accessible immediately when the card is powered up—in fact, it will often be transmitted to the terminal or host system as a part of the card's response when it is powered up; this is known as the answer to reset (ATR). But the other data are only accessible after the card has received a security code, either from the cardholder or from the software application, which matches a code held secretly in the card. Often there will be several such protected areas: One part may be accessible after the cardholder has given a code, another is unlocked by a program controlled by the card issuer, and a third part is only accessible to the manufacturer. Sometimes the codes themselves are held in a further area, which is only accessible once, at the time when the card is issued or personalized. Once the codes are written, a fuse is blown to prevent them from being altered.

Protected memory cards are available with memories from under 100 bits to several kilobits. Memory sizes are most often quoted in bits, as these units are most commonly used in counting applications, such as telephone cards and vending. They can also be used for simple access control with a single numeric key or PIN.

Basic protected memory cards are inexpensive ($0.50–$5, even in hundreds) and can be bought and programmed by a fraudster using almost any PC-connected smart card reader. Assuming that he or she has access to the memory mapping and that there is no further encryption in the system, a counterfeit card which will appear to the system like a real card can be produced. The data on a valid card are protected by the PIN and other security codes, but these are subject to the limitations of all single-code schemes: Fraudsters can often find ways of obtaining legitimate PINs through carelessness or deception.

Secure logic

A further group of memory cards not only controls access to memory, but also restricts the ways in which the external application can write to or read from the memory. For example, prepaid telephone cards can contain logic to prevent the value from being increased. They may also contain special logic that enables the application to authenticate the card cryptographically by holding a *hashed* form of the card data (see Chapter 5), which alters as the value is decremented.

Secure logic chips are also likely to include several of the special security features described in the next chapter, including irreversible programming, transport codes, and logical address scrambling. Many also contain specific provision, known as *antitearing*, to protect the data against corruption if the card is removed prematurely or if power is removed from the card before it has completed the transaction.

Most modern telephone cards, public-transportation cards, and pre-payment cards follow this pattern. They are considerably more secure than the basic protected memory cards, and if the correct issuing procedures are followed (see Chapter 11) the risk of counterfeit cards should be virtually eliminated. It is also important that the authentication mechanisms built into the cards are used. Otherwise it could still be possible to design a card or system that would respond correctly, even if it were not a true copy. (In the first edition of this book, we mentioned the possibility of designing a card that always gives a "true" output, regardless of the input, and this method was subsequently used by Serge Humpich to produce his "Yes-card," which resulted in his being jailed for defrauding the French bank card system.) Designers of smart card security schemes using this type of card must be able to implement a basic public-key encryption program, as described in Chapter 5 and in several of the references in the bibliography.

Microprocessor cards

Development

For maximum security and true portability of data, the card must incorporate a microprocessor. With a microprocessor card, the data are never directly available to the external application: They must always pass through the microprocessor, which can carry on a dialogue with the application.

In the first smart cards, the microprocessor and memory were separate. One of the early Bull patents involved incorporating the two into a single chip, and nowadays several IC manufacturers are able to produce specially designed chips that meet all of the requirements of security, power, and memory capacity within the required thickness of chip.

Conventional

Most smart card microcomputer chips adopt a conventional *von Neumann* architecture, as shown in Figure 7.1. As well as the processor, memory, and power control, there are special circuits for the security and for communication with the outside world. Within the card, data are passed through a bus, under control of the security logic.

The partitioning of the microcomputer's memory, and the logic that governs access to the partitions, is central to the security of a microprocessor card (although, as we will see in Chapter 8, most smart card micros also make use of other techniques to prevent unauthorized reading, counterfeiting, or masquerade). All access to memory therefore passes through the security logic. The most important areas of memory—containing the card's own secret keys—are not accessible from outside the card at all.

Many smart card micros also have an arithmetic coprocessor, which helps particularly with cryptographic functions. Some also have a random number generator, which is used when two-way authentication is required.

There is a single input-output port, either a bidirectional serial contact or a wireless interface. And every smart card has some form of power-control and clock-control circuitry. This varies in complexity quite considerably; where the card issuer has little or no control over the environment

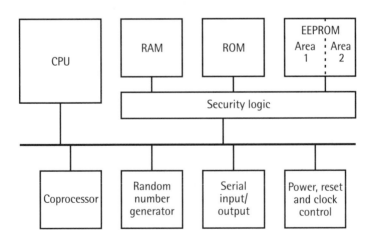

Figure 7.1 Smart card microcomputer architecture.

or readers in which the card will be used, the card must be prepared for a wide range of clock speeds and voltages.

From a security point of view, virtually any smart card microprocessor can be regarded as impossible to counterfeit. Legitimate cards or chips can be bought, however, or they can be stolen at any stage in the manufacturing or personalization cycle, and this could allow data to be copied onto those cards. The procedures for card manufacture, personalization, and issuing must ensure that cards are never in a state where keys or confidential data could be extracted or where stolen cards could be programmed with apparently valid keys.

State change

Smart cards using the conventional architecture described earlier are reaching current limits of speed and memory capacity. These limits are set by the power generated by each transistor junction, which must be dissipated through the chip, and by the maximum physical size the chip can be before it becomes too fragile. These limits are gradually being extended, but a solution that meets the requirements of some applications is already available.

Instead of using a conventional microprocessor, which executes microcode through a single set of registers, programmable gate arrays (PGAs) have a large number of transistor junctions wired to perform logic according to the *program*. The output depends only on the state of the inputs (which may include memory devices); thus, they are called state or *state-change* devices rather than microprocessors. PGAs, which occupy a smaller chip area, have been programmed to emulate popular microprocessors and will often be faster at given tasks.

Field programmable gate arrays (FPGAs) can also have user programs added to the logic, and are therefore almost as versatile as the original devices.

PGAs and other state-change devices have the added advantage that they are not in general subject to the patents and license conditions of the microprocessor manufacturers. Although this has become less of an issue since the expiry of the main Innovatron patents, manufacturers of state-change devices have been freer to develop their own solutions free from license requirements.

Against these arguments, state-change devices are easier to analyze and emulate than microcontrollers with their programs. They are typically made in smaller volumes and do not benefit from the special techniques and controls that are applied to the higher end smart card microcontrollers.

Cryptographic

A further group of smart cards is used for applications that require the card itself to perform specialized cryptographic functions. As we saw in Chapter 5, many of these functions require arithmetic on very long operands, for which most general-purpose processors are too slow.

Such cards therefore incorporate arithmetic coprocessors, incorporating special functions which not only allow multiple word arithmetic (up to 1,028 bits in some cases) but also special algorithms for solving common cryptographic problems.

Crypto cards also incorporate the full range of security devices described in Chapter 8, such as memory-address scrambling, auto-detection of hacking attempts by multiple PIN tries, power-circuit manipulation, or even electron microscopy.

They represent the pinnacle of commercial microcomputer and system security technology at the beginning of the twenty-first century. Although in most cases these cards are not provably secure in a mathematical sense (we cannot demonstrate that no modes of failure that would compromise security exist), they do appear to defeat in a practical sense every mode of attack currently envisaged. Several teams of cryptographers, physicists, and engineers have set out to break them using massive computing resources, but so far all of the modes considered involve very low probabilities of success (see Chapter 8 for more details of possible attacks and countermeaures).

Contact and contactless

Contact cards

Until now, we have generally been considering cards that will be used as replacements for magnetic-stripe cards, principally in financial and access-control applications. Most of these cards will make their interface with the outside world through a set of six to eight contacts, as defined in ISO 7816 part 1. All early smart cards, in Europe at least, were of this type and met one of these standards.

The contacts themselves are a potential point of weakness in a smart card system:

- The contacts can become worn through excessive use or damaged by a defective reader or in a pocket;

- The leads from the microcircuit to the contacts are of necessity very thin and can break or become detached when the card is bent or otherwise stressed;

- They represent an obvious starting point for any attack;

- The contact set in the reader is a mechanical device, which can break or be damaged either accidentally or maliciously. Chewing gum and cyanoacrylate adhesives are both popular tools for vandals.

Contactless cards

To overcome these problems, manufacturers have developed contactless cards, in which the microcircuit is fully sealed inside the card and communicates with the outside world through an antenna wound also into the thickness of the card. Power may be provided by a battery, but the preferred solution (adopted by most current designs) is for power to be collected by an antenna, which also normally acts as the communications antenna. The antennas may be quite small (less than 1 cm diameter) or wrapped around the outside of the card, but they can collect sufficient power from the electromagnetic field to power the IC as well as providing the communications path. Up to 5 mW can readily be coupled in this way, which is adequate for most single-function cards.

The systems can be tuned so that the reading distance is large (up to 50 cm) or small (the card must be placed in contact with the reader). However, this sensitivity is mostly a function of the reader rather than the card, so cardholders could, in theory at least, be justified in fearing that transactions could take place on their cards without their knowing it.

The ISO standards for contactless cards (ISO 14443) are currently driven by the two main manufacturer groups, with Type A cards representing the MiFare licensees and Type B the Motorola group. Final agreement on these standards has been delayed in order to ensure manufacturer independence. A third type (Vicinity cards, with a reading distance of up to 50 cm) is covered by ISO 15693.

The transportation industry, particularly public transportation, is a major potential user of smart cards. Applications in this industry depend in nearly every case on the use of contactless cards, which offer a much higher transaction rate per terminal because the card does not have to be inserted and removed. Contactless cards and the associated readers are also more reliable, require less maintenance, and should have a longer life than contact cards.

Against this range several arguments, most of which are commercial rather than technical.

* The speed of communication for a contactless card depends on the power of the reader/transmitter and on the reading distance. Transaction times must be kept short, as the card may move out of range very quickly. Communication speeds are now much higher for contactless cards than for contact (115 Kbps rather than 9.6 Kbps), in order to minimize this problem.

* Some contactless cards are slightly thicker than the ISO standard, because they will rarely be placed into a reader and the extra thickness makes them more robust. Such nonstandard cards are becoming less common.

* Although in principle any smart card microcircuit could be used in a contactless card, in practice the number of manufacturers is small and the range of card types available is limited. More recently, 32-bit cards suitable for multiple applications have become available in contactless form, which should help to overcome this problem.

* The banking industry in particular has reservations about any card that can carry out a transaction without being inserted into a terminal, since the cardholder may not be aware of "covert transactions."

* Contactless cards are currently more expensive than corresponding contact cards, largely because fewer cards have been made. This difference is being eroded rapidly by increasing volume and may shortly disappear.

For all of these reasons, contact cards are still preferred for the majority of applications. Nevertheless, the market for smart cards in public transportation is now a major driver for growth in the market, and several other industries are looking to public transportation as the initial "anchor" application which could justify the initial cost of card issue. This will lead those other industries to port their applications onto contactless cards as well.

Combi cards

For consumers, the most frequently quoted advantage of the smart card is its ability to handle multiple applications on a single card, thus reducing the number of cards in people's wallets and the confusion this can cause. For some applications, the ability to share cards is also necessary because of the high cost of cards and the need for a network of readers. But if one group of applications requires contact cards and another requires contactless, the two may be unable to share a card.

There are two solutions to this problem; both overcome some of these problems but require that we can read the same card using either contact or contactless reading.

Pouches that will accept a contact card and effectively convert it into a contactless card are available. The pouch makes contact with the card's contacts, but it has its own power and communications circuitry driven by an antenna. The combination has the speed limitation of the contactless card, but any contact card can be used. This would, for example, enable a bank card also to contain a public-transportation application.

While pouches are certainly convenient, they offer from a security point of view the least satisfactory answer: The contact point of weakness still exists, whereas now many more data are being passed between card and reader by radio (which can be intercepted). Current designs do not allow the card to tell whether it is being addressed by radio or by contact, and a radio reader could therefore carry out a transaction that should have been restricted to the contact mode. Pouches should be seen as a temporary solution suitable for lower security applications.

True combination cards have both contacts and an antenna. Some designs have two independent microcircuits, one addressable by the contacts and the other by the antenna. This is the most secure, but also the most costly solution in the long term.

Other designs allow the same microcircuit to be addressed by either method. Either dual-ported memory is used or the same microprocessor supports both contact and contactless interfaces (see Figure 7.2). In this case it is essential that the logic knows which route has been used and controls access to memory accordingly. Some applications may only allow one form of access (contact or contactless), or even may not wish the other to know of the existence of the application. This *dual interface* design is now the preferred architecture.

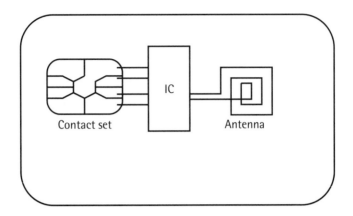

Figure 7.2 Dual interface combi card.

A German manufacturer has claimed a patent on all such designs, but from an engineering perspective such a broad-ranging claim seems unrealistic.

Form factors

The very term *smart card* implies a card-shaped device, and we have come to take it to mean a credit card size. However, smart card technology is used in several applications where other shapes and sizes are preferred.

Modules

A smart card module (also known as an ID-00 card) is more or less the smallest size of plastic card that can accommodate a smart card chip and its contacts (see Figure 7.3).

Modules are used as security application modules (SAMs) within readers and terminals; a SAM contact set can be mounted directly onto a printed circuit board and in some applications the module is locked in place by *potting* the relevant parts of the reader (fully enclosing them in a resin compound that cannot be removed without destroying the circuitry).

They are also used in mobile telephony, as subscriber identity modules (SIMs) in global system for mobile communication (GSM) telephones and in some portable instrumentation.

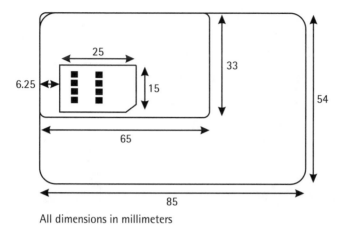

All dimensions in millimeters

Figure 7.3 Module and minicard sizes.

Minicards

Minicards (see Figure 7.3) are a size between the module and the ISO 7810 credit card. The slightly smaller size is potentially cheaper to manufacture, and the card is more rigid and less liable to break due to mechanical stress.

Minicards are being proposed for several public-transportation applications, where they often correspond to the size of existing magnetic-stripe tickets. Because of this, there has been some difficulty in agreeing on an international standard for minicards, with each country arguing for the size of its existing tickets.

Keys

Smart card chips can also be built into key-shaped devices (see Figure 7.4). These are physically very robust and can also be made with a mechanical action in addition to their logical functions. The contacts appear either in a row or as a series of rings around the shank of the key.

USB devices

Smart card–like devices can be attached directly to a USB port connected to a PC (see Figure 7.5), and are likely to be used for authentication or for access to secure data. Such devices must use proprietary communication protocols rather than the ISO standards, but may be attractive for single applications where interoperability is not an issue. Because there is no need for a separate reader, the cost per installation is very low.

Figure 7.4 Smart key. (Photograph courtesy of Nexus (G.B.) Ltd.)

Figure 7.5 USB token. (Photograph courtesy of Aladdin Knowledge Systems.)

Others

Many other shapes and sizes are also possible, provided only that the carrier is large enough to support the microcircuit. Many of these proprietary devices gain security from their proprietary nature: They will usually be uneconomic to reproduce or to hack into. For high-security applications, however, they should be treated with some caution, as none of these devices incorporates the specialized techniques or years of development that have gone into the high-end products in the standard shape and size. Speed, memory capacity, and special features are all sacrificed in the majority of proprietary designs, and the smaller manufacturers are also unlikely to have the extensive controls and procedures in place to protect the manufacturing and personalization process fully.

8

Chip Card Security Features

Carrier

As with magnetic-stripe cards, the basic material of the chip card is polyvinyl chloride (PVC) or a similar thermoplastic. PVC is most commonly used for international bank cards because it is cheap and can be embossed. Acrylonitrile butadiene styrene (ABS) is also widely used; this slightly more brittle plastic cannot readily be embossed, but it will withstand higher temperatures and thermal cycling. This allows the chip to be bonded to the card at a higher temperature (with PVC cards the module is usually glued into the card). Polycarbonate also has a higher temperature specification and is used in mobile telephone cards, which call for operation at up to 100°C. Polyethylene terephthalate (PETP), which offers a very flexible and light card, is used widely in Japan.

The card production process will be discussed in more detail in Chapter 11; however, with all of these materials the chip and its contacts are normally embedded in the card in the form of a module. The module is bonded into a milled or molded indentation in the card body, usually after the card has been printed.

Several manufacturers have variations on this process, each with its claimed advantages. One manufacturer uses polycarbonate as the carrier material in a shortened card production process; the chip is inserted directly into the card and the contacts are printed onto the surface of the card using a conductive ink as a part of the printing stage. Another manufacturer has shown that it is possible to make cards using wood as the carrier material.

No card material is inherently superior to all others, although some may be better for individual applications. Card security is improved slightly when the chip or module cannot be removed from the card after manufacture (this would make some forms of circuit analysis easier); with resin-bonded modules removal is sometimes possible.

What is important, however, is the life of the card. It is the mechanical features of a smart card that determine its life; failure or unreliability usually arises from contacts becoming damaged or the links from the chip to the contacts being broken. This can come about as a result of mechanical stress (particularly twisting) or thermal stress (differences in expansion coefficients between the plastic and the wires). In applications where high reliability is required, a plastic that deforms little with heat and is not prone to fatigue failure with repeated twisting should be used. Of the widely used materials, ABS and polycarbonate probably meet this requirement best.

External security features

Smart cards can incorporate several of the security features described in Chapter 4, including:

- Microprinting;
- UV-sensitive ink;
- Hologram;
- Signature strip or printed signature;
- Cardholder photograph.

These features will often be less relevant in the case of a smart card (which is designed to be authenticated electronically) than for a magnetic-stripe card, for which visual inspection is usually the only form of card authentication.

Bank-issued smart cards will also often incorporate a magnetic stripe so that they can be read in countries or by retailers that are not equipped with smart card readers. It is a rule of the international card schemes that a retailer who accepts, say, Visa cards must accept all valid Visa cards, not just domestic cards.

Cards may also carry a small bar code in one corner; this is a batch control used in manufacturing and personalization, and does not form part of the operational characteristics of the card.

Chip

The key component of the smart card is the integrated circuit embedded in it. As we described in Chapter 7, this may be a memory, protected memory, microprocessor, or FPGA chip. Several semiconductor manufacturers now make ICs specifically designed for smart cards, generally using:

- *Low-power CMOS technology,* operating at voltages between 5V and 1.8V (chips operating down to 0.8V and below are being designed).

- *Very small feature size* (the width of the smallest line or separation on the chip): 0.5μ is now common, and the most advanced IC production lines are now working at 0.15μ. Smaller sizes still are in view.

- *Special manufacturing controls,* described further in Chapter 11.

These chips are normally available from the IC manufacturer in the form of sawed wafers, tapes, or modules for direct embedding in cards.

Microprocessor

Most current smart card micros are still 8 bit, usually based on Motorola 6805 or Intel 8051 designs, and with clock speeds of 3.5 or 5 MHz. Manufacturers are now turning to 32-bit word sizes for their more powerful chips, in particular for use with multiapplication operating systems. Sixteen-bit processors are less common.

In order to increase processing speed, higher clock speeds are required. It was found that if the speed of the external clock was increased significantly, RF interference effects were experienced. In addition, changes to the ISO standard governing clock speed negotiation ran into difficulties, and so manufacturers have moved towards generating their own clock onboard, synchronizing from time to time with the external clock. This allows them to operate at much higher speeds—20 MHz is already used on the fastest chips.

The latest stage of development is the introduction of reduced instruction set (RISC) microcontrollers, which offer high performance and low power consumption. Some manufacturers are also moving to pipeline processing and instruction sets optimized for direct execution of Java code. The effect on transaction speeds, particularly where authentication is a major function, is very significant: Typical figures might be a reduction from 3 seconds to 0.5 seconds. The limiting factor is now often the communication speed between card and terminal, which is still set at 9,600 bps in most instances.

Memory

Memory is the largest variable item in the design of a smart card IC. The memory is first divided into areas according to the type of semiconductor memory:

- Read-only memory (ROM) is used for storing the fixed program of the card, also known as its *mask*. The mask is sometimes thought of as the card's operating system, but often it is the only program. Part of the ROM memory (known as user ROM) may also be available for what we can think of as application programs running in the card. ROM is efficient in both space and power requirements.

- Programmable read-only memory (PROM) is used for loading a card's serial numbers or other fixed values. This part of memory is usually very small (up to 32 bytes) and accounts for little space or power.

- *Flash* memory is sometimes used for storing additional programs for execution within the card, as well as for storing blocks of data which will be read and extracted all at the same time. This is because flash memory can be written using the normal card communications interface, but it can only be erased as a single block. Flash memory is more efficient in space and power than E^2PROM, but less so than ROM. It is not widely used in smart cards today, although it is becoming more common.

- *E^2PROM* is used for most data storage—the smart card's equivalent of the hard disk on a PC. It can be read or written at any time (subject to the security logic), but is very costly in space and in power. It is nonetheless the most important determinant of the card's flexibility of function, and so today's cards are being provided with up to 64 KB of E^2PROM.

For many smart card chips, the E^2PROM accounts for a large portion of the total chip area, as shown in Figure 8.1. Two important parameters for the life and reliability of chip cards are the number of write cycles the E^2PROM can reliably accept and the data retention period (the period for which the memory will reliably retain data). Although the quoted figures for these are usually quite high (typically 10,000 cycles and 10 years, respectively), both are statistically determined (they depend on the rate of migration through the semiconductor junction) and in a large card population some cards will start to show unreliability well before the quoted limits.

♦ Random access memory (RAM) is used for temporary working storage. Data in RAM is lost when the card is removed. Whereas PCs and other small computers normally have access to tens of megabytes

Figure 8.1 Smart card microcomputer chip. (Photograph courtesy of Infineon.)

of memory, smart card micros typically have memory sizes from 128 to 1024 bytes.

◆ Ferro-electric RAM (FRAM or FERAM) is starting to be used in smart cards. This type of memory consists of RAM with an additional layer, which has the effect of making it nonvolatile (i.e., it retains its memory without power). It can therefore be used in place of E^2PROM, over which it has two major advantages:

- Write time is the same as read time: nanoseconds compared with milliseconds for E^2PROM;

- The power required for writing is also much less than for E^2PROM; this is particularly important for contactless cards because it reduces the total power requirement of the chip.

- FRAM has a much higher packing density than E^2PROM and can perform as volatile or nonvolatile memory. This makes it possible for a manufacturer to design the whole of memory as a single block, designating areas as ROM or RAM at customization time.

- FRAM was for many years limited by the tight control maintained by a patent holder; it has now been developed independently by other manufacturers and is embedded in smart card chips now becoming available in 2001. Memory sizes of 256 KB and higher are envisaged using FRAM.

Different applications require varying proportions of these different memory types, and it follows that cards are divided into applications according to these ratios, as well as by their power and the functions built into the mask.

For example, a card with a large flash memory may be used for storing a history of operations, for storing health records, or for telemetry. Applications such as bank cards require a large amount of ROM because of the complexity of the programs, whereas the new generation of security modules for mobile telephones requires very large E^2PROM areas to store subscriber profiles as well as passwords, phone books, and call history.

Coprocessors

Smart card chips that are designed for use in security applications very often have a built-in coprocessor. In some cases this processor carries out standard multiple-length arithmetic operations (multiplication and exponentiation) in hardware; in other cases the coprocessor is designed to execute common cryptographic functions, such as DES encryption or RSA signatures, directly.

Today's cryptoprocessors will perform RSA 1,024-bit signatures in under 150 ms and authentication in under 100 ms. DES encryption times are tens of microseconds, which makes them almost negligible for transaction-based applications. These chips do not come cheap, however: 32-bit cards incorporating fast cryptoprocessors cost at least five times as much, volume for volume, as 8-bit cards with comparable memory sizes.

Memory management

The memory of a smart card chip is controlled by memory-management circuits, which provide hardware protection against unauthorized access. This protection makes use of a hierarchy of data files: The highest level is termed a master file (MF), below this may be several layers of dedicated files (DF), and finally one layer of elementary file (EF) (see Figure 8.2). Access to a lower level must pass through each of its parent files in turn; this forms a logical channel to the elementary file. Parent files contain the access rules and control information relating to the next layer down. ISO 7816-4 (see Appendix A) defines the security mechanisms appropriate to provide both authentication and confidentiality of the data, keys, and control fields

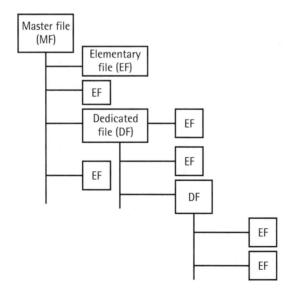

Figure 8.2 Smart card file structure.

during normal access. The card itself provides these security mechanisms, and the designer must ensure that all fields are correctly identified and that the structure of the file hierarchy is appropriate for the application.

Input–output

All microprocessor smart cards make use of a single bidirectional serial input-output interface. As we will see, the electrical connection may be through contacts or by electromagnetic coupling. The standard for this communication is set by ISO 7816-3, which defines a character- or block-oriented asynchronous protocol (known as T = 0 or T = 1, respectively).

The speed of communication is set by a combination of the clock speed (defined by the terminal device) and the dividers in the card. The terminal "wakes up" the card by applying power followed by a reset (RST) signal. The card responds with an answer to reset (ATR), which tells the terminal what type of card it is and which communications protocol it will use.

The input-output interface for a contact card consists of two connections: I/O and GND, supported by an optional clock signal.

Chip security features

We have already mentioned the visual security features on the card. The security of the chip itself starts with the design process: In laying out the circuits, the designers of security-conscious microcontrollers have a number of tricks available to make the analysis of their devices more difficult. They include:

- *Layering:* A complex microcontroller will consist of many layers of logic. In normal circuit design, it is regarded as good practice to keep like with like; to deter analysis, however, functions can be spread across several layers, in a seemingly random fashion. Elements regarded as vulnerable to analysis (particularly ROM) will be buried as far as possible in a lower layer.

- *Bus scrambling:* Buses may not be laid out in sequential order and may be in a different order on different layers.

- *Address scrambling:* Memory may not be laid out in the same order as its logical addressing would suggest.

- *Dummy components and functions* may be added where space allows.

- *Active components or links* may pass through or over the final *metalization* and *passivization* layers, which protect the chip against atmospheric effects and must usually be removed for any intrusive analysis.

The chip design may also include several special functions for the detection of attempted analysis or hacking, such as:

- *Low- and high-frequency detection:* This detects attempts to operate the chip very slowly for analysis or to introduce short pulses (glitches) which might induce errors.

- *Temperature detectors:* These can shut down the chip when it approaches its maximum temperature. This is as much for reliability as an antihacking feature.

- *Light detection:* Detects exposure to light after removal of the nontransparent epoxy resin covering the chip.

- *Etching detection:* Detects removal of the *passivization layer* by using etching acids.

Few if any chips actually have all these features. All cost money, and most smart card schemes are sensitive to the cost of the cards, so many of these features would typically only be available on relatively expensive cards (costing $5 or more).

There is also a strong emphasis on security during manufacture of both the chip and the card. Physical security at the plants is carefully controlled. The manufacturer's identification number and a serial number are written to one-time programmable (OTP) memory, which "locks" the data. The chip is tested before it leaves the foundry; if it passes, the fusible links that were used to perform the testing are blown, and flags are set in a security register.

This process is repeated after the chip is incorporated into a card and may be further repeated after prepersonalization, as described in Chapter 11.

Contacts

A normal (contact) card has up to eight contacts, the position and designation of which are defined by ISO 7816–2 (see Figure 8.3). Up until 1999, many French cards used a different contact position, designated in the ISO

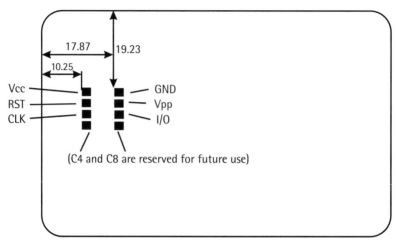

All dimensions in millimeters

Figure 8.3 ISO 7816–2 contact positions and designations.

standard as "transitional." All cards now use the standard position, shown in Figure 8.3.

The contacts may be made of gold or another conductive material; they are usually connected to the chip itself by very thin wires during the module manufacturing process. Because of the flexibility of the card, these connections are a potential source of unreliability. The contacts themselves have a limited life, particularly when they are used with cheap readers using *sliding* contacts.

Antenna

In a contactless card, all input-output—and normally the power—is transmitted by radio-frequency (RF) signals rather than through contacts. A coil antenna is built into the thickness of the card, either around its circumference or with a larger number of turns around the chip itself, within the size of the module.

Some early contactless cards used a battery within the card. In today's cards, the RF-control circuits in the chip pick up sufficient energy through the antenna to power the chip itself and to communicate with the read-write unit. The signal uses frequencies designated for telemetry: 125 kHz or

13.56 MHz. The card communicates by modulating the RF signal, using one of the standard modulation types: amplitude modulation (AM), frequency modulation (FM), amplitude shift keying (ASK), frequency shift keying (FSK), or binary phase shift keying (BPSK). Card designers try to use the system that will require the least power; a protocol such as code division multiple access (CDMA), which would yield the most secure communication, requires too much power for today's card antenna/chip combinations.

Power requirements drive contactless designs; typical microprocessor cards require 5–8 mW of power, while 1–1.5 mW may be adequate for small protected memory cards. The power required increases as a cube of the distance from card to reader. More power can be transmitted at the lower frequency, but data rates are lower. With the lower frequency, the distance from the card to the read-write antenna can be up to 1m, with the higher frequency the maximum is usually closer to 20 cm. The maximum data-transfer rate, however, increases with frequency (more than 100 Kbps for the 13.56-MHz cards).

In many applications, the actual reading distance is kept smaller than the maximum possible. The larger the reading distance, the more likely it is that there will be several cards in range at one time; thus, the chance of communications collisions is greater. In financial applications (even where a bus fare is being deducted from the card) users like the more positive action of bringing the card into contact with the reader, rather than having value removed from the card while it is in their pocket. For other applications (building-access control or season tickets in public transportation) the ability to read at a distance is a positive advantage, and for these applications some form of anticollision protocol is required. These protocols have been designed for a wide variety of networks and do not pose any problems to the system designer.

Contactless smart cards are able to make use of any of the security mechanisms (controlled memory access, card and reader authentication, data encryption, etc.) available to contact cards. Some of the lower end memory products do not make use of these mechanisms, and they therefore raise the specter of data being read from a card, or a transaction effected, without the knowledge of the cardholder.

Where card and reader authenticate each other, the only question is whether a second reader could gain any useful information from passively listening in on an exchange between a valid reader and the card. This is probably more of a theoretical possibility than a practical likelihood, but if the eavesdropping device had knowledge of the public keys of the two devices, there is a possibility that it could extract a session key (or any other symmetrical key) from the exchange and thereby decode data encrypted using that key. For systems requiring high security, therefore, key distribution within a contactless card scheme should only be carried out in a secure physical environment.

Mask

The fixed program of a microprocessor card (often referred to as its operating system) is held in ROM, and it is called the card's mask. Different versions of a card can be produced on the same microprocessor chip by using different masks. Unlike a general-purpose computer, for which the operating system only carries out such functions as file management, scheduling, and intertask protection, the mask of a smart card often performs application functions: reducing the value in a purse, for example, or comparing a digitized signature with a stored pattern. Some of the application logic may be contained in the mask: A smart credit card, for example, will not complete a transaction until the terminal has executed the correct sequence of checks.

Smart card microprocessor manufacturers can supply a default mask, including the minimum functions required to make use of the card's features. This is often built into the development kit for the card. Most smart card manufacturers (which take the module and build it into a card) have developed their own masks for popular applications, as have several card personalization companies (which are principally concerned with those elements that can be loaded after the card is manufactured).

As the mask must be programmed into the chip when it is manufactured, its security forms part of the manufacturing process. For high-security chips, the manufacturer insists on a tight control of the development tools and process as well. These processes will be described in Chapter 11.

From a security point of view, a microprocessor card operating system consists of a set of layers, as shown in Figure 8.4. The innermost layer is concerned with handling access to individual fields and how those fields may be manipulated. The file manager deals with the DF-EF structure and the access rules for those files. The command manager interprets commands and checks whether that command is valid at this point in the

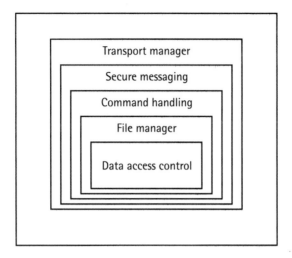

Figure 8.4 Card operating system structure.

operation. The next layer implements the secure messaging defined in ISO 7816–4, which is designed to provide both authentication and confidentiality of the data exchanged between reader and card. Last, the transport manager provides the low-level communications functions required by ISO 7816–3. This structure provides a very high level of confidence in the ability of the card to protect the data stored in it, under all normal operating conditions.

Attacks and countermeasures

There are two broad classes of attack on chip cards: *external attacks*, which operate the card from outside (through its normal contacts or radio interface) and *internal attacks*, which involve performing operations directly on the chip.

An internal attack requires the micromodule to be removed from the card, normally by dissolving the plastic using acetone. The epoxy resin is then removed in an acid bath. Once the chip is exposed, it may be analyzed using a light or electron microscope, probed using needles, or modified using a focused ion-beam (FIB) system. The capacitor network which generates the E^2PROM programming voltage may be destroyed using an ion beam, so that the data are not altered during successive operations [1].

Most of the techniques involved require facilities that are available in semiconductor laboratories in several universities; they are used by manufacturers to test their own and their competitors' products. Designers are therefore able to take into account the ease of any of these attacks and the likelihood of their being successful.

Internal attacks

Examples of internal attacks include the following.

Microscopic analysis

Light and electron microscopes can be used to inspect the physical structure of the chip, and this could permit the analysis or reverse engineering of the device, or of specific interesting areas such as the security mechanisms or data storage [2].

To protect against this examination yielding any useful information, the buses may be scrambled or the logical memory structure may be mapped onto physical memory in an apparently random way: one bit is stored here and another there, so that without access to the memory mapping the analyst cannot reconstruct the data. Some randomization is used on all but the simplest smart card microprocessor chips. More complex chips are constructed in layers, with the memory-management functions, PROM, and other highly sensitive elements in the lowest layers.

To protect further against electron microscopic analysis, security chips may include a *metalization layer* over the whole chip; this in turn includes some links into the main circuitry so that any attempt to remove the metalization layer destroys the chip. One manufacturer uses a further layer of silicon as a security layer. Some high-security chips also include circuits that can detect the radiation of an electron microscope and destroy all the data in memory (E^2PROM data are likely to be erased by the beam in any case).

Mechanical probing

When the chip is exposed, the analyst gains access to several input and monitoring points which are not brought out to external contacts. Very sharp needles—mechanical microprobes—can be used to make galvanic contact with the metal tracks on the chip. The data transported over the buses of a device in operation can be tapped, although there is a risk of damaging the chip. Chips manufactured using 0.5μ or smaller technology, or protected by a metalization layer, can generally not be probed using today's tools.

Use of test modes

During card manufacture, test modes and test contacts are provided on the chip. They are disabled (by software and/or hardware) once the tests have been successfully completed. Sometimes these features can be located, reestablished, and then misused to retrieve secrets from the card. Again, antietching layers are the best protection.

External attacks

External attacks involve the use of a card—often a large number of cards. The attack can be designed using an old or invalid card, and then carried out on a live card. The chances of success for most attacks is quite low, but if sufficient resources and time are expended, the chances of a successful attack by one of these means cannot be ruled out. Where the motivation, whether financial, political fanaticism, or intellectual challenge, is high enough, there will be instances of sustained and coordinated attacks, and even the finest encryption system is subject to the risk that the first key combination attempted is the right one.

Some external attacks include the following.

Circuit analysis

Circuit analysis involves low-frequency manipulation of the inputs and observation of the outputs. Many chips include detection circuits, which detect any attempt to operate the chip outside its normal clock or timing parameters and can shut the chip down. The use of onboard clock generation also blocks this type of attack.

Measurement of operating parameters

Smart card chips are quite slow compared with other computers. It is therefore possible to measure the time it takes to perform certain tasks and make deductions concerning the results. For example, a "hit" in a search table will often produce a shorter operation time. Or the current can be measured, yielding information about the quantity of data being written. Millions of repeated observations of this type can yield useful data for analysis. Some chips therefore introduce null operations, or random variations within a single operation, to destroy the usefulness of any information that may be gained by external measurements.

One form of parameter measurement which has been shown to be successful is *differential power analysis* [3]. DPA measures the variation in power used by the chip as it performs arithmetic (particularly cryptographic functions, which can easily be identified), and analyzes these variations to extract keys and other data. With knowledge of the RSA algorithm, for example, DPA can extract a private key from many cards. As with most such attacks, DPA can readily be thwarted by specific design, for example ensuring that any coprocessor runs in parallel with, but decoupled from, main processor operations [4]. But private keys stored in cards without such protection are vulnerable.

Deliberately induced errors

It has been proposed that data, particularly the secret keys of a public key cryptogram (see Chapter 5), could be extracted by inducing an error in an iteration (preferably the penultimate) and observing the result of the last iteration. This could be done by introducing a glitch on the clock or power supply or by radiation. Although the chip should ideally cease to operate as soon as its operating range is exceeded (as it would in the case of a clock or voltage variation), very few chips carry detectors for radiation.

In practice, however, this type of error could not be introduced systematically, and the observer would therefore have to analyze an exceptionally large quantity of data with no certainty of obtaining a result—a level of effort that would be disproportionate to the rewards in virtually all cases.

The standard protection against this attack is for the smart card to authenticate the terminal before commencing any operation [5]. The chip can additionally protect against such attacks by performing operations twice or by reversing a cryptographic transform and checking that the original value is obtained. Although these lead to increased operation times, they will provide the necessary security in those applications in which it is paramount.

Reliability factors

All of the elements described here have an effect on the reliability of the card. The carrier must, as we have seen, have very good strength and flexibility, without imposing undue stress on the chip itself or on the connection to the contacts. The contacts themselves must be thick enough to last the life of the card, bearing in mind the range of readers in which it may be used.

The chip size is very largely dependent on the area required for E^2PROM, as this takes up the most space. It is usually argued that a smart card chip should not exceed 20–25 mm^2; above this size the chip itself becomes too fragile. The amount of E^2PROM which can be accommodated in this area depends in turn on the feature size, which is linked to the operating voltage. Current smart card technology, with feature sizes of 0.35μ–0.7μ and voltages of 2.7V or 3.3V, can accommodate up to 16 KB or so of E^2PROM within this area.

Although greater complexity will in principle reduce the reliability of the card, in practice the security and other design features used in the more complex cards result in a unit that is at least as reliable as its simpler brethren.

Sample card specifications

Tables 8.1 and 8.2 give some examples of card and chip specifications for different applications. The security features mentioned are examples only; all cards will have several features not mentioned here. Manufacturers will not generally give a full list of all security features.

TABLE 8.1   Sample Card Specifications

Manufacturer	Card	Application	Processor	Memory	Crypto Functions	Other Security Features
Orga	OPPC	Prepaid telephone card	—	104 or 237 bits	Off-line & on-line authentication (Eurochip functions)	Decrement-only counter
Oberthur	AuthentIC	Internet authentication	8-bit	1–4 KB	RSA or ECC, DES, 3DES, DSA	Common Criteria EAL 4/5
Gemplus	GemShare D/C	Debit card	8-bit	2 or 8 KB	3DES	EMV static data auth, PIN verification & management. EMV approved.
Schlumberger	Cyberflex SIM	GSM SIM card (phase 2)	8-bit	8 KB E²PROM	COMP128, A5/1 (GSM algorithms)	ITSEC E4; JavaCard authentication
Philips	MiFare Prox	Dual-interface card for ticketing & payment	8-bit core	64 KB ROM, 2.3 KB RAM, 16 KB E²PROM	3DES coprocessor; optional 32-bit public key cryptoprocessor for RSA, DSS, ECC	Multiple sensors & security barriers; hardware memory management
Motorola	MV5000	Dual-interface card for transportation and e-purse	8-bit	32 KB ROM, 8 KB E²PROM	Coprocessor for DES, 3DES	DPA resistance, secure messaging; direct support of Proton e-purse

TABLE 8.2   Sample Smart Card Chip Specifications

MANUFACTURER	CHIP	TYPICAL APPLICATIONS	PROCESSOR	MEMORY	CRYPTO FUNCTIONS	OTHER SECURITY FEATURES
Hitachi	H8/3114	Electronic banking, Multos	8-bit	32 KB ROM, 2 KB RAM, 16 KB E²PROM	Coprocessor for DES/RSA	Primitives for EMV, Multos, and high-speed peer-to-peer communications. ITSEC E6
Atmel	AT90SC	Satellite TV and other streaming decryption	8-bit RISC	16–64 KB Flash, 16–64 KB E²PROM	16-bit crypto-processor for exponentiation and DSA	Clock speed & intrusion detection, secure layout, supervisor/user modes
Infineon	SLE66CL160S	Dual interface chip for bank cards, electronic purses	16-bit nonstandard architecture	32 KB ROM, 16 KB E²PROM	DES, 3DES, ECC	Multiple sensors, security PROM, secure layout, ROM code hidden
NEC	V-way 32	Multiapplication, JavaCard	32-bit RISC	128 KB ROM, 3 KB RAM, 32 KB E²PROM	Cryptoprocessor for RSA, DSA, DSS	Active security circuits, DPA shield
ST Microelectronics	SmartJ	Multiapplication, JavaCard	32-bit RISC	8–128 KB E²PROM	Direct execution of Java bytecode and public-key instructions	On-chip clock generation, DPA protection, scrambling, hardware memory protection

References

[1] Anderson, R., and M. Kuhn, "Tamper-Resistance—A Cautionary Note," *Proceedings of Second Workshop in Electronic Commerce*, USENIX Association, Oakland, CA, 1996.

[2] Blythe, S., et al., "Layout Reconstruction of Complex Silicon Chips," *IEEE Journal of Solid-State Circuits*, Vol. 28, No. 2, 1993, pp. 138–145.

[3] Kocher, P., "Differential Power Analysis," *Risks Digest*, ACM Committee on Computers and Public Policy, Vol. 19, June 1998.

[4] Chari, S., et al., "Towards Sound Approaches to Counteract Power-Analysis Attacks," *Crypto 1999*, University of California, Santa Barbara, CA: Springer-Verlag, 1999.

[5] Carlson, R., "Authentication: Smart Cards Are the Next Killer App," Corporate publication, Authentic8, Doncaster, VIC, Australia.

9

Multiapplication Operating Systems

Since the early days of chip cards, commentators have suggested that one of their main advantages lay in the possibility of reducing the number of cards in our pocket. (For example, see [1].) Because the applications would share some common data, they could interact, allowing, for example, any application to make a payment using the payment function on the card, or collection of loyalty points associated with the payment. A transport card could include not only a bus pass, but also some small change, a telephony function, and perhaps even storage for some airline tickets and a passport. As we will see in Chapter 18, this raises a large number of commercial and operational issues, few of which have been fully resolved to date.

But to take advantage of the flexibility of the chip card, and to some extent also to offset its high cost, developers need to find ways of allowing multiple applications to reside on a card.

There is also a need for chip card applications to be platform-independent, and to allow high-level languages to be used in their development. Throughout the 1980s and 1990s chip card applications have been developed in assembler or in C, and have been specific to a chip and hard-mask combination. This has tied card issuers and scheme operators to one card type, and discouraged suppliers from developing general-purpose programs that could be used across a wide range of card types and schemes. Development times have been long: 18 months is probably a typical lead time for a product to reach the market. This is not acceptable in most of the commercial environments where chip cards are proposed, and indeed it does explain why so little progress has been made in some areas over the last five years.

In these fast-moving commercial environments, new products and services are developed and launched in as little as four months. This is much shorter than not only the development time for a low-level chip card application, but also the life of the card. So it must be possible to load and update applications on the card after the card has been issued.

These issues are addressed by the family of products known as multi-application operating systems. In order to manage a card base using multiapplication operating systems, some further development tools and management tools are also required.

Objectives

For a card to support multiple applications, the first requirement is for the applications to be truly independent of one another. This means that applications must not be able to overwrite each others' data, nor read the data without the other's agreement. When one application runs, it must not affect the operating environment for the other in any way. And the same is true when a new application is loaded to or unloaded from the card.

This implies some form of "firewalling" between the applications. The key element of this is memory management: Memory is allocated to authorized applications under control of the operating system. This means that there must be some way to authorize applications: Authorize them to be loaded, to run or to use RAM and E^2PROM storage areas.

A multiapplication operating system will also allow the use of high-level languages in development. In particular, mainstream programming methodology is based on an "object-oriented" construction and languages that can be compiled and run on many different platforms. That is an advantage for chip card systems, because the portability of the card means that it may be used in many different terminal types, ranging from PCs to dedicated terminals and in-vehicle systems or even to set-top boxes and microwave ovens.

User authentication is a necessary part of any application authorization scheme. Generally the OS will not specify a particular form of authentication (e.g., PIN or biometric), but it should provide features to support user authentication, and to link that authentication into the application management structure.

Multifunction versus multiapplication cards

Multiapplication is not necessarily the same thing as multifunctionality. A single application may have several functions—what a programmer would recognize as different *modes* or *methods* of the object. For example, a university campus card could have a student union card, a library ticket, a computer-room access card, and a record of attendance at lectures. These could all be driven from a single application on the card.

Even where there are several separate programs on the card—perhaps a payment and a home-banking application—the card does not need a multiapplication operating system if the programs are developed and tested together, and will always be present when the card is issued. But if we envisage updating one or the other of the applications after the card has been issued, or if one of the applications is not completely trusted by the other, then a multiapplication card is required.

Using a multiapplication card allows applications to be developed, tested, and updated independently, but it also brings a need for inter-process protection and for managing shared data. And even where we can place a high degree of reliance on the card operating system, extensive testing is still strongly recommended to avoid any operational or technical incompatibilities.

The use of dynamic loading and unloading (after the card has been issued) increases flexibility but the processes and environment where this takes place must be controlled.

Multiapplication cards have been designed to allow several application issuers to share a card. This aspect will be addressed further in Chapter 18, but here we are mostly concerned with applications issued by the card issuer—the application issuer and the card issuer are the same organization. (Nonetheless, companies implementing multiapplication card management schemes are recommended to allow for multiple application issuers as well.)

Functions

Most of today's smart cards do not have a true operating system in the sense that larger computer systems use the term. There are no external resources to control other than the I/O port, and only one application runs at a time. In the longer term, there is no reason why this must remain true; with faster processes and more memory, applications could be suspended or run in contention for the I/O port, cryptoprocessor, or other functions contained within the card. This would require a more conventional operating system with all of its overheads.

With current card masks, programs are not held in the same memory as data; interprocess protection is thereby made much simpler than with, for example, a PC system. It is easier to ensure that one application does not affect the integrity of another application; the operating system must simply ensure that only one application can gain access to its data, with the exception of data that is explicitly shared.

This is where the problems begin. Defining a mechanism that allows one application to share its data with another authorized application, probably in specific ways (e.g., application A creates the data field, B may write and read the data, C may only read it), requires an operating system designed for that purpose.

Where, in addition, the data fields are protected by keys, the key management of the card must be extended so that applications can manage their own keys. (In most smart cards, there is one key file, which is used to give access to all data fields.) It may even be necessary for two applications (but not necessarily all applications on the card) to have access to the same key. This requires a comprehensive scheme for granting access to an elementary file (EF)—or possibly even to fields within that EF—according to the application and access control conditions to be met.

Such a scheme involves considerable overheads both on data storage and on processing power, and so it can only be implemented on microprocessors at the top of today's memory and power range. However, there are now several cards—usually 32-bit cards with 16 KB or more of memory—which meet these requirements and can be used to support the operating systems described later in this chapter.

Products

In 2001 there are three main contenders in the multiapplication operating system marketplace: JavaCard, Multos, and Windows for Smart Cards. Each has its merits and, as we will see, they are aimed at slightly different markets and applications.

JavaCard

JavaCard was developed by Integrity Arts, a California company subsequently bought by Sun Microsystems.

JavaCard leverages the advantages of the Java language—what Sun calls "Write Once, Run Anywhere."[TM] Java programs are used in embedded devices and PCs, up to top-end NT and Unix servers. The language was designed for use across networks, and many of us are most familiar with the Javascript language which can be embedded in Internet pages and interpreted by browsers. It produces relatively efficient, self-protecting code. One of the normal development tools is a "sandbox" in which Java applications are run during development and debugging.

JavaCard offers a subset of Java (many Java commands would not be appropriate in a card with no external resources), which can in principle be run in an 8-bit card with 8 KB of E^2PROM. In practice, most JavaCard implementations are on 32-bit cards with 16 or 32 KB of memory.

The JavaCard definition is an application program interface (API), which must be implemented by the card manufacturer. It will normally be overlaid on a proprietary operating system, although some chip manufacturers are now designing chips which will execute Java byte code directly, making a direct implementation more likely. The current version of the JavaCard API (version 2.1.1) offers interprocess protection, a mechanism for loading and updating applications, and a basic security framework.

There are several JavaCard implementations already commercially available. They are most advanced in the telephony sector, where the experience suggests that effective compatibility and interoperability between cards and applications from different manufacturers should be achievable during 2001.

JavaCard offers the advantages of using a mainstream programming language, an API, and toolkits that are license-free and downloadable [2].

Multos

Multos was developed by the same group which created the technology for the Mondex electronic purse. Maosco, the company which runs it, is now owned by a consortium of card schemes, software companies, and card manufacturers.

Multos is partially integrated with the proprietary operating system, rather than being overlaid on it as in the JavaCard case. It executes (by interpretation) a language called MEL (Multos Executable Language);

applications can also be compiled from C using a Maosco-supplied development kit. Maosco has worked with card manufacturers to produce approved implementations.

Multos' strength lies in its security architecture. As a result of its origins in the banking community and in high-security cards, it focuses on very tight control by the card issuer of everything that happens to the card. Every application load or unload has to be certified by the issuer, using a load or unload certificate provided by Maosco. Inevitably, this high-security infrastructure imposes some overheads, and Multos cards will appeal most to those issuers for whom maintaining control is a high priority, whereas others may focus more on cost per application or on performance.

Windows for Smart Cards

Windows for Smart Cards was developed by Microsoft, and its focus is on extending the Windows environment down into the chip card. Thus, developers with experience of developing Windows applications (typically in Visual Basic) will be able quickly to start writing programs for chip cards.

Because the Windows 2000 operating system incorporates chip card support, not only for secure log-on authentication but also using the Kerberos public key infrastructure [3], an increasing number of PCs will be shipped with integral smart card readers—typically using either the PC/SC or OCF standards, which will be described in Chapter 10. Application developers will seek to make use of these devices whenever they have applications requiring authentication or small amounts of portable and personal data storage.

Windows for Smart Cards incorporates a security structure that extends the Windows file management system down into the card. This is unlikely to be considered adequate for high-security applications without some user-added features, but will meet most PC-related requirements.

Unlike the other two OSs described here, Windows for Smart Cards is implemented directly by the chip manufacturer and replaces the proprietary operating system on the card. In fact, the chips are developed specially for use with this OS. (See Table 9.1.)

Windows for Smart Cards' advantage lies in the Microsoft developer base, most of whom are already using some of the relevant tools, and all of whom now receive a quarterly set of CDs with the latest development tools relating to chip cards. (At the time of writing, Microsoft is still trying to

TABLE 9.1   Multiapplication Operating Systems Compared

	JAVACARD	MULTOS	WFSC
Application load management	Issuer choice (can use Open Platform)	Maosco	No standards at present
Application language	Java subset	MEL	Visual Basic (or Visual C)
Virtual machine	MicroJava VM	Multos	WfSC API
Card operating system	Proprietary	Partly integrated	Integrated
Chip	Standard ISO 7816	Standard ISO 7816	Specially designed

resolve a negative antitrust ruling in the U.S. courts, but this is unlikely to affect its smart card offering directly.)

Open Platform

Open Platform was originally promoted by Visa (as the Visa Open Platform) but it has now been merged with other JavaCard-related initiatives and is now run by a consortium called Global Platform, which includes telecommunication companies and technology companies as well as Visa and other banking associations [4].

Open Platform takes the JavaCard concept further, and adds some of the extra features likely to be required by card issuers, in particular control of the personalization process and of application loading and unloading. It introduces card issuer, personalization company, and application issuer security domains. But it is not as prescriptive as Multos, and issuers can determine their own levels of control. Global Platform has said that it intends the Open Platform to be usable with Windows for Smart Cards as well as with JavaCard.

Open Platform has been used for GSM SIM cards (see Chapter 12), credit and debit cards (see Chapter 14). There is also a corresponding set of specifications relating to terminals, which enable the construction of applications independent of the environment in which those applications will run.

Security

A key feature of a multiapplication operating system is the level of security it offers, in particular the degree of protection between applications. Both

Multos and Open Platform have published protection profiles under the *Common Criteria* (see Chapter 3), which give some assurance that these aspects have been designed into the product, and Multos has a certificate in this respect under the older ITSEC criteria.

At present it appears that Multos has the edge on provable security, partly because its rigid commercial structure helps to enforce that security. The principles it embodies are popular with banks and other high-security applications, although it has not yet completely lost its alignment with the Mondex/MasterCard grouping, which may limit its take-up across the spectrum.

JavaCard, and in particular Open Platform, implementations should be able to reach the same level of security, but will need to put in place or use an appropriate certification infrastructure in order to achieve it. JavaCard is currently stronger in those applications which cross sectors (e.g., telephony and banking).

For applications where the key driver is compatibility with PC systems (and this could include e-commerce), Windows for Smart Cards could be the preferred vehicle, although this proposition is not yet as well developed as those for JavaCard and Multos.

References

[1]   Ognibene, P. J., "Smart Cards: The Ultimate Personal Computer," *Miami Herald*, October 2, 1988.

[2]   www.java.sun.com/products/javacard.

[3]   Schiller, J., J. Steiner, and C. Neuman, "Kerberos: An Authentication Service for Open Network Systems," *Usenix Winter Conference*, Dallas, TX, November 1988.

[4]   www.globalplatform.org.

10

System Components

A smart card is an element in a distributed computing system. Having no user interface of its own, it can only operate as a part of a system. The security of any system is only as strong as its weakest link; thus, even if the smart card itself were totally impregnable, the data could be compromised by other parts of the system or by the processes and procedures that make use of the card. This chapter is concerned with the other system elements, and in Chapter 11 we will look at the procedures.

Reader

Chip card readers are more accurately known as card-accepting devices (CADs), because they can write to the card as well as read data from it. Other terms used include read-write unit (RWU), write-read unit (WRU), or interface device (IFD).

Most systems designers and users will want to purchase complete reader assemblies or terminals meeting one of the common interface standards, such as PC/SC, EMV, or GSM.

Manufacturers of other devices, such as set-top boxes, kiosks, or instrumentation, may have a need to purchase contact sets and interface chips, but this option is likely to lead to a higher priced product and should only be used where there is a need for a specific interface or form factor. Designers of systems requiring motorized readers, or combinations of chip card and magstripe, must also become more involved with the detailed issues, some of which are addressed in this section.

Contacts

The card reader for a standard contact card consists of a set of contacts, some logic or intelligence that acts as an interface between the card and the rest of the terminal, and the card transport mechanism.

The basic element of this reader is the set of between five and eight contacts which establish the electrical connection with the card (see Figure 10.1). These contacts may be sliding (the card is moved into place under the contacts) or landing (the contacts only close onto the card once the card is in place). Sliding contacts have a slightly higher chance of making a good connection, but at the price of a greater insertion pressure and greater wear on the card contacts. It is also more difficult to ensure that power is removed from the card immediately if the card is withdrawn; if this is done too slowly there is a possibility of giving wrong results or of

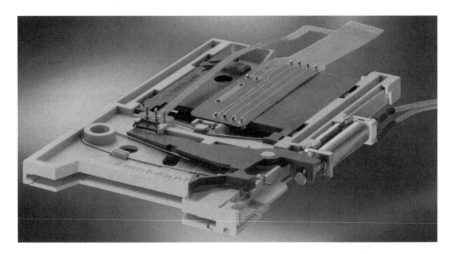

Figure 10.1 Chip card contact set. (Photograph courtesy of Amphenol-Tuchel Electronics GmbH.)

feeding the wrong voltages to the other contacts. The best contacts have a small amount of sliding as they land.

Card transport

Readers for smart cards may be of the manual insertion type or motorized. Both have implications for system reliability and security:

* *Insertion readers* (see Figure 10.2) are much cheaper: They have few moving parts and require very little power. They can, however, be vandalized (by inserting a metal plate, screwdriver, chewing gum, or even fast-setting cyanoacrylate adhesive). Insertion readers are generally unable to capture cards; card capture is less important for a pure smart card (because the chip itself can be disabled by the reader), but most bank cards will continue to have magnetic stripes as well for some time. And there is a higher risk of the card being withdrawn before the transaction is completed; as we will see, this requires specific logic or software to bring the card and system back to a known and consistent state.

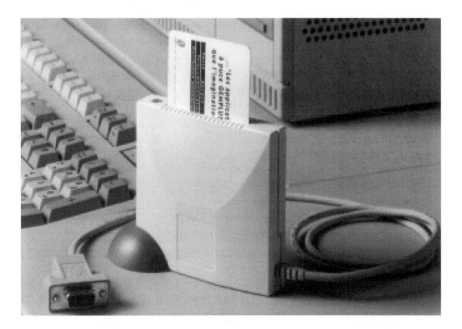

Figure 10.2 Manual insertion reader. (Photograph courtesy of Gemplus.)

* *Motorized readers* (see Figure 10.3) are able to avoid the vandalism problem to a large extent by using a shutter (although on many existing readers this is opened by detecting the magnetic stripe). They are able to capture cards, and the card is fully under the control of the system while the transaction is in progress; this leads to a more consistent operation. On the other hand, motorized units require more power and, having more mechanical parts, must be made more robust. Hence, it is considerably more expensive to achieve a given reliability.

Control electronics

There are now several commercially available chips for handling ISO 7816-3 protocols. These handle the initial power-up, clock, and answer to reset sequences, and convert standard asynchronous or synchronous data at TTL levels into the T = 0 and T = 1 protocols and levels required by the card. Some versions will also handle limited numbers of error conditions.

Some of these chips are designed to be driven by a microprocessor, in which case the micro can undertake some of the command interpretation

Figure 10.3 Motorized reader. (Photograph courtesy of Omron Electronics, Ltd.)

and even integrated authentication and encryption functions. Here we are getting into the realm of the integrated terminal; the dividing line between readers and terminals is not fixed.

In integrated terminal applications, however, particularly where security and reliability are critical design factors, there may be a need to incorporate three further sets of features, which are not handled by the standard ISO 7816-3 chips.

- Positive control of the card transport status: Not only must power be applied when the card is fully inserted and when it is removed, but in many instances we want to send messages to customers: "Please insert card fully," or "Remove card now." In many applications designers want the card to be inaccessible to the user while the transaction is in progress: It is released by a solenoid when the transaction is complete. In these cases there must also be a means of releasing the card mechanically if the power fails during the transaction.

- The ability to send PINs or other CVM data securely to the card, without risk of interception, and to display messages to the customer which can only have come from or through the card. This requires a secure physical unit: The reader itself and the authentication electronics must be tamper-proof, or at least tamper-evident. This is usually done by *potting* the electronics in an epoxy resin and ensuring that if the housing is opened, the memory is erased.

- If the communication between card and PINpad is remote, then there exists an ISO standard (ISO 9564) for encrypting the PIN, but this does not cover any outwards communication from the card, nor other forms of data.

- Where keys must be stored securely within the reader, or where data must be signed using a private key for authentication (see Chapter 5), a SAM is often the best way to protect the key. A SAM is a smart card module, usually incorporating a cryptoprocessor and memory, which forms part of the reader electronics (see Figure 10.4).

Contactless card readers

Card readers for use with contactless cards are able to dispense with the contacts and card transport; in place of these they make use of an antenna,

Figure 10.4 Reader electronics and security application module. (Photograph courtesy of Landis & Gyr S.A.)

which is usually wound as a coil and mounted behind a plate. The reader can be mounted in whichever position is likely to be most convenient for users and is likely to give the best coupling with the card. For some applications it may be located at waist height, to allow the card to be carried in a pocket, but other operators prefer to have users bring their card into contact with the reader.

In addition, the electronics must control and decode the RF signal. This will often require logic to detect multiple cards in the field and to ensure that communication only takes place with one card at a time. Again, there are now commercially available chips or modules that will perform these functions. As we described in Chapter 8, this logic is not as secure as that used by the best military and commercial radio systems, but as we will see in Chapter 16, it meets the requirements of public-transport operators around the world.

Terminal

Card-accepting devices are built into many forms of terminals—free-standing devices that connect to a network. These terminals may take several forms, such as PINpads, PC readers, electronic point-of-sale (EPOS) or electronic funds transfer at the point-of-sale (EFT-POS) terminals, ATMs, vending machines, and access control. Today we can even regard the mobile phone as a portable chip card terminal.

PINpads

A PINpad is a small handheld terminal comprising a keypad, display, and smart card reader (see Figure 10.5). Its purpose is to allow the cardholder to enter a PIN without having to disclose that PIN to the shop assistant or to anyone else. PINpads therefore often have some form of cover over the display, and sometimes even the keypad, to stop others from observing the PIN as it is entered. The design should take into account the physical layout of the environment in which it will be used: Would it, for example, be possible for someone to be looking over the shoulder of the person entering

Figure 10.5 Secure PIN pad. (Photograph courtesy of Hypercom Financial Terminals.)

the PIN ("shoulder-surfing")? Since most PINpads conform to a standard layout, someone using binoculars or a telephoto lens can observe the operation from afar.

The PIN must also be transmitted securely to the card; if the card reader forms part of the PINpad, then the whole terminal can be made tamperproof. System designers should ensure that if the terminal is opened (for example to insert some additional wiring to monitor the operation), it is not subsequently possible to reactivate the terminal without its passing security controls.

Where the card reader does not form an integral part of the PINpad, the PIN must be encrypted as it is passed from the pad to the card.

PC reader

Smart cards are increasingly used to give access to computer systems, and data must often be loaded onto cards using a PC. In the world of electronic commerce, as we shall see in Chapter 13, smart cards are beginning to play an important part in providing secure payments across public networks. For all of these purposes, smart card readers must be attached to PCs.

Most PC readers conform to one of two sets of standards.

- ◆ PC/SC: the PC Smart Card specifications were originated by Schlumberger, Bull CP8, HP, Microsoft, and Siemens Nixdorf. Their aim was to allow chip card readers and chip card applications to be added to PC systems just like any other peripheral or application. PC/SC readers may use a serial or USB port, PS/2 keyboard connection or a PCMCIA (PC card). Figure 10.6 shows a typical device.

  PC/SC provides a structure and a set of interfaces for card and reader manufacturers and operating system suppliers to work to in designing software drivers and application program interfaces. PC/SC implementations are generally inexpensive—many of the devices are cheap enough to be distributed free or as part of a package by service providers, and they do make it very easy for application developers to use chip cards. But PC/SC is a relatively loose specification and many of today's implementations lack the sophistication and features that will in the long term be required if users start to depend on their chip cards for a wide range of security functions (for example, several serial-port implementations lock out other serial devices). A list of PC/SC-compatible devices is on the workgroup's Web site [1].

Figure 10.6 PC/SC reader. (Photograph courtesy of Towitoko Electronics GmbH.)

- OpenCard Framework (OCF) [2]: these specifications were originally developed by IBM for set-top boxes, but have been adopted as
  a reader standard by the Java community. OCF is a more tightly
  defined interface, and can be used on Windows PCs running C or
  C++ applications as well as within Java systems. But it has not
  equaled PC/SC for penetration in the Windows world.

For applications requiring a secure PIN entry, there may be a need to
integrate the reader with a keypad and display. Again there are two possible
approaches:

- The EU has sponsored a development by the French and German
  payment card associations called FINREAD. This has defined a
  chip card reader for PC use, with integrated keypad and display (see
  Figures 10.7 and 10.8). The PIN never passes outside the reader and

Figure 10.7 FINREAD card-accepting device. (Photograph courtesy of Xiring SA.)

Figure 10.8 FINREAD card-accepting device. (Photograph courtesy of CPS Europe.)

Fax Back Card

Free information on other Artech House books...

To ensure that you stay up-to-date on the newest Artech House books of interest to you, fill in the address details below so that we may add you to our mailing list. Please fax or mail it to one of the locations listed on the other side of this card.

Name: _____

Position: _____

Company: _____

Address: _____

_____

E-mail: _____

(If you would like to receive new title information by e-mail.)

Please indicate your areas of interest:

☐ Telecommunications
Wireless
Networking

☐ Computing
Software Engineering
Security/E-Commerce
Database Management

☐ Microwave

☐ Radar
Remote Sensing
Electronic Defense

☐ Antennas & Propagation

☐ Signal Processing

☐ Optoelectronics

☐ Technology Management/
Professional Development

AH Artech House Publishers BOSTON • LONDON

only data from the card is displayed. This is being used by the French Cyber-Comm project (see Chapter 14).

* The chip card reader can be incorporated into the PC keyboard itself (see Figure 10.9). Since the keyboard is essentially a one-way (input) device, communication with the chip card must be handled through a special driver, and a separate serial-port connection unless the universal serial bus (USB) is used.

EPOS or EFT-POS terminal

Most large retailers in northern Europe and North America now make use of electronic point-of-sale (EPOS) terminals. In addition to the functions of the standard electronic cash register, the EPOS terminal is connected to a network and can be programmed for additional functions. The cash register has access to a database of items in the shop; as goods are sold, their bar codes are scanned and the purchase is recorded. Full details of the transaction are sent to a back-office computer, and selected data are sent to the head office for further analysis.

Where smart cards are to be used for recording the payment or customer details alongside the transaction, a smart card reader or PINpad must be attached to the register or terminal. EPOS manufacturers have not yet developed fully integrated functions for smart card reading; thus, at this

Figure 10.9 Keyboard with integrated chip card reader. (Photograph courtesy of Cherry Keyboards GmbH.)

stage it is likely that any reader must have a sufficiently high level of intelligence to carry out all the smart card–specific functions; the EPOS terminal acts as a network component.

In the case of card payments, suppliers have found that with current terminals and interfaces, better performance can be achieved by putting all the authentication and cryptography functions into the card-accepting device than by splitting them between the CAD and EPOS terminal. This also makes it easier for organizations such as the card schemes to certify these devices.

The downside of working this way is that more functions are performed in hardware or firmware, so it is less easy to update applications than if most of the functions were performed in the EPOS terminal software.

Card readers and terminals to be used primarily for payment functions in a retail environment must conform to the Europay-MasterCard-Visa (EMV) standards defined by the major international card schemes. EMV, which is described in more detail in Chapter 14, is a more closely defined specification than ISO 7816, and includes a certification process. Card-accepting devices must meet the electrical and protocol specifications known as Level 1, while in order to be certified to EMV standards a terminal must also implement one of the card schemes' transaction-flow specifications (known as Level 2). Devices with this certification will be considerably more expensive than basic Level 1 or PC/SC units. Lists of EMV-certified units are held on the EMV Web site [3].

ATM

The next form of reader is the automated teller machine (ATM). The issue here is to provide smart card reading as well as magnetic-stripe reading, as a single upgrade. ATMs are always connected on-line to the bank network to perform transactions and include symmetric data encryption in their software. But the smart card subsystem must be able to perform the card authentication (an asymmetric algorithm) and other intimate functions before passing the transaction to the host system. These functions are now available for most ATMs commonly used in Europe, although at the time of writing few of them are actually certified to EMV standards.

Vending machine

Many vending machines have already been adapted to accept smart cards; the operators find that the benefits of reduced vandalism and not having to collect cash far outweigh the extra cost of the card. Most of these machines use proprietary cards and interfaces, and their security depends not only on the conventional measures discussed elsewhere in this book, but also on the control exercised by the manufacturer over the technology.

As smart cards become more pervasive, vending machine operators and their customers will want to make use of stored value cards issued by others, including electronic purses offered by banks. The vending operators will have to adapt their systems to accept cards meeting the open standards; many have already started to move in this direction.

Most vending machines accept only one form of payment. There is usually only one operation, which also lends some security to the system. Typically, insertion readers are used, and the transaction data are stored in flash memory until it is collected by a host system. It is not usually necessary to store details of the card used, although provided that the data are stored in an encrypted form there are few security implications in storing the card number as part of the transaction.

The use of smart cards will permit increasingly complex and higher value transactions to be performed through kiosks and vending machines, including ticket issue for public transportation and events, sales of holidays and insurance, and ordering for home delivery. In these cases, it is likely that a bank-issued card will be used, and the kiosk must incorporate the full facilities required for card authentication, cardholder verification, transaction encryption, and secure storage until the unit is polled by the host system. Many will also need to be able to connect on-line to a bank host to authorize the transaction amount.

As with other retail terminals, EMV certification is an issue if the kiosk is to be used for payment transactions using bank-issued cards.

Access control

Smart cards are not yet widely used for access control. There are several reasons for this:

- The action of reading a smart card is slower and more awkward than swiping a magnetic-stripe card;

- The extra cost is rarely justified unless other cardholder-verification techniques (biometric or PIN) are used at the same time;

- Access-control companies have concentrated their development on on-line systems.

Contactless smart cards are now becoming more widespread, and are starting to replace the traditional Wiegand tag in many security-conscious establishments. Smart card readers for access control usually consist of aerial pads located adjacent to doors, although in some military or high-value financial situations the extra safety of a contact card combined with a biometric check will still be preferred.

Others

Free-standing smart card terminals, often similar to a PINpad but with additional intelligence and memory, are used in a variety of applications, including retail-loyalty schemes, gaming control, and off-line functions such as loading and balance checking in closed-loop systems.

Stored-value schemes often require simple balance readers, which can be made available to the cardholder. These are usually in the form of a key-fob, just large enough to take the corner of the card and to display a value on a single-line LCD display.

Multicard operations (e.g., transferring money from an electronic purse onto a membership card to pay a subscription) require some form of electronic wallet, and in some models of electronic purse this will also allow value to be exchanged between individuals. The wallet looks like an electronic calculator, with the addition of a smart card reader slot.

Terminal protection

Most terminals must be physically protected. If they are not in a secure area, then we may need to consider regular polling or dialing in, and the terminal should incorporate self-diagnostics so that it can inform the central system of any malfunction or attempt to tamper with it.

Chip card terminals, particularly kiosks, are now able to incorporate web servers, to allow remote users to view the terminal parameters and performance, and e-mail clients, to send messages in the case of alarms or malfunctions.

Network

The role of the card

The card, as we have said, is only one element in a distributed computer system. In some cases, it is providing data to the system; in other cases, it allows access to a program or to data on the host system. Most card operations are transactions—the card and system exchange some data, agree that they are prepared to do business with one another, exchange details of the transaction, and then break the connection. Usually there is only one exchange of data between the card and the host—all the other functions are performed by the reader or terminal.

There is one special case in which the card "transaction" has a much longer duration, and this is when the card is used as a key to gain access to a computer system. In this case, the transaction lasts for as long as the on-line session and the card should remain in the reader for all of that time. Log-on systems often fail to log the user off if the card is removed!

Network security checks

Where a card transaction must pass, for any part of its journey, through a network that cannot be regarded as fully secure, precautions must be taken to ensure that the purpose of the card—be it authentication, confidentiality, integrity, or any combination of these—is not compromised.

We must start by assessing which parts of the network are secure. We have to look at:

- *The system design:* Do we control all of the network? What protocols are used and do they provide the security we need? Are there backups if critical items fail? Are the backups equally secure? What other access points to the network exist?

- *Physical security:* Is the building secure? Is all critical equipment in a room with controlled access?

- *Organization:* Who has access to the system (at any level)? Who controls that access? Who has responsibility for critical elements? Does the staff policy meet the needs of security? If part of the network is subcontracted, does the contract include the security requirements and appropriate service level clauses?

These aspects are covered in more detail in the references in the bibliography.

Provision of network security

For those parts of the network that we cannot control, we must implement the security measures ourselves. For *reliability* and *integrity*, this will mean providing backup paths, message numbering, and message authentication checks. For *confidentiality*, messages must be encrypted; the keys themselves must be exchanged by an alternative route or under cover of a master key that is rarely changed. For *authentication,* a hashed value of all messages must be signed and the signature checked by the recipient; this process should ideally be carried out in both directions.

The best way to provide these security functions in a complex network environment, or when parts of the network cannot be completely trusted, is to use end-to-end authentication. For most card systems, this means that all messaging between the card and the ultimate host system is covered by a single authentication and encryption scheme. The principle used in the SSL/TLS protocols (see Chapter 13) usually provides all the security services required, although to keep the system manageable the number of top-level certification authorities must be kept to a minimum—preferably not more than two or three.

Fallback and recovery

The network design must also take into account the need for fallback and recovery. Failures will occur, and some of these will happen during a transaction. When this happens, the two ends of the system often end up out of synchronization, with the data updated on the host system but not yet on the card or vice versa. The design of a smart card scheme must provide a method for canceling a transaction that has not been completed. This is fairly simple for a single-stage transaction, but can be quite complex for operations that require changes to several different datasets.

One method is to use a common logging and storage area, accessible by all parts of the system; it remains the responsibility of each component to use the information in this logging area correctly. Alternatively, systems can be designed using the "two-stage commit" procedure preferred by database designers—any changed data is written to a temporary record, and only when completion of the transaction has been confirmed does the temporary record replace the main record.

Host systems

It is not the place of this book to enter into the details of the security provided by host systems or of the measures which can be taken to improve that security. These subjects are covered in more detail in several of the references in the bibliography. In general, mainframe-based hosts already have most of the general security measures appropriate to systems of this type; smaller UNIX or Windows NT hosts are usually technically sound but often suffer from inadequate procedures and staff controls. In a smart card environment, however, several additional areas need attention.

These mostly concern encryption and key management. Data that must be kept confidential should be stored either in encrypted files or files with formal access rights and regular security reviews. Keys must be stored in a secure manner: There are various techniques for this, but the most common method for secret keys is to divide them into components, with a different member of staff responsible for each component. They are then entered via a special device, which recombines them to form the key.

These devices, which are known as host security modules (HSMs), come to form an important part of host system security (see Figure 10.10). As we shall see in Chapter 14, their use is obligatory in financial systems. They can perform encryption functions both on-line, under command of the host, or off-line, via an attached terminal or PC. Key elements are held in semiconductor memory in tamperproof compartments. If they are tampered with or the power is removed, then the keys are destroyed.

Once a secure communication has been established between two parts of the system, keys can be securely stored on the host, encrypted under a key which is known only to the HSM. They can be transmitted to other parts of the system using the secure communications path, and operations such as digital signatures and signature checking are performed by the

Figure 10.10   Host security module. (Photograph courtesy of Zaxus Ltd.)

HSM itself. Modern HSMs can perform functions such as RSA encryption or signature verification many times faster than software implementations, and they can directly implement encryption functions required by applications such as card payment.

Many smart card systems are concerned with personal data; in this case the implications of data protection on the data stored must be kept in mind. In some countries, it may be necessary to encrypt all data to avoid unauthorized access, while in others the data must be stored in such a way that they can readily be accessed by cardholders and independent inspectors.

References

[1]   www.pcscworkgroup.com.

[2]   www.opencard.org.

[3]   www.emvco.com.

11

Processes and Procedures

Many processes are involved in the lifetime of a smart card; this chapter considers the security implications of each and how they are normally addressed.

Any system is only as strong as its weakest link, and nearly all computer system security breaches involve bad procedures or equipment breakdowns rather than failures of the security mechanisms themselves. Poor system design, or failure to enforce correct procedures, is the equivalent of installing an expensive lock and then leaving the key in the door.

Many systems designers are technologists and forget the importance of properly designed and documented *procedures*. The move towards ISO 9000 approvals has gone some way towards raising awareness of this very important subject, but too many organizations feel that because they cannot achieve ISO 9000, quality management is not important.

The fundamental requirement is to monitor the status of the card at every stage in its life cycle. In this chapter we identify 12 or more states within the card's life cycle, ranging from the early stages of manufacture and test to expiration and destruction; the number of states relevant in any given case may be higher or lower than this (see Figure 11.1). The card-management system must have provision for all the relevant states.

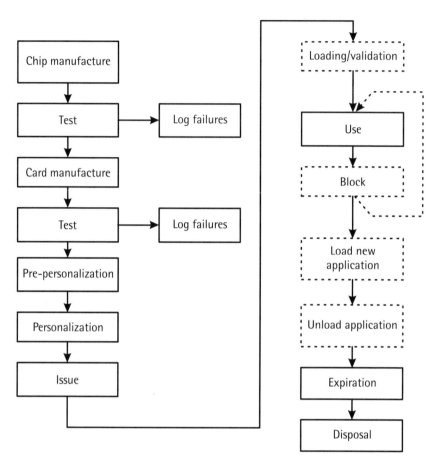

Figure 11.1   Card life cycle.

Chip design

The security of a chip card starts with the design of the chip itself. For commercial reasons, all design of large-scale integrated circuits is carried out on secure computer-aided design (CAD) systems in facilities with tight access controls.

First, the design is expressed in terms of design requirements (for different types of memory, I/O, and security). This is laid out in block diagram form and input to the CAD system. Further details of each block are then supplied to the CAD system, which as with all chip design starts to produce circuit layouts in the form of layers of materials.

A smart card chip, whether memory or microprocessor, consists of a very large number of interconnected transistors. The layers represent p-type and n-type silicon, metal and oxides (insulators), with connections between the layers. The design and layout of a smart card microcontroller are subject to intense scrutiny for possible security weaknesses. After several iterations of this process, the design is complete and ready to be committed to photolithographic plates.

Manufacture

The manufacturing process also starts with the chip (see Figure 11.2). The semiconductor manufacturer produces wafers of silicon carrying a large

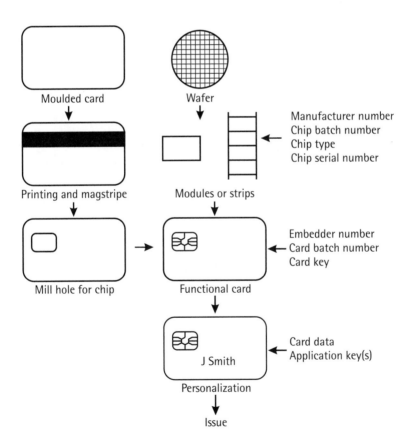

Figure 11.2 Typical smart card production process.

number of chips. These are divided into individual chips, tested, and then mounted either on *modules* (small plastic cards including the contacts and very thin wire connections from the chip to the contacts) or on strips, depending on how the next stages of manufacture will be carried out.

Different smart card manufacturers use slightly different processes, often as a function of the card material they are using. Figure 11.2 shows a typical process, in which a standard PVC card is manufactured, printed, and has its magnetic stripe applied. A small hole is then milled into the card surface, and the module is then inserted into this hole and glued in place using a strong epoxy resin. The advantage of this method is that the chip itself does not have to pass through any of the earlier stages, which can involve higher temperatures and heavy stressing of the card.

The card itself is then tested and passed to the personalization stage.

Because complete smart cards are almost impossible to counterfeit, it is important to control theft of cards and components throughout the manufacturing stage. Again, strong emphasis is placed on good physical security in the premises in which chips and cards are manufactured. The resin glues used to bond the chip into the module are designed to be stronger than the silicon underneath, so that any attempt to remove the chip should destroy it.

The chip also passes through several clearly defined stages, and it is usually designed so that once it passes to the next stage it cannot go back. During manufacturing and testing, links used to provide access to testing functions or to load serial numbers and keys are blown after use.

During the initial mask loading, the semiconductor manufacturer (known as the *foundry* in the trade) writes data to the chip, which cannot be changed by subsequent operations. For maximum security, this data will include the manufacturer number, batch, or wafer number, and component type. Once the chip has been tested, it is given a component number, and the *fabrication lock* is then blown before the chip is passed to the smart card manufacturer (sometimes known as the *embedder*).

The card manufacturer embeds the chip in the card and often adds a further level of functionality (the soft mask) to the card, specific to the application. This process is also called *prepersonalization* or sometimes *customization*. The card is tested again, and this time the embedder identification number, batch number, and card number are written to the chip. A card-specific key may be added at this stage. The *prepersonalization* process is completed by blowing the relevant lock, so that all of the data entered can never again be altered.

The card and chip have by now passed through several stages of testing. Products that fail any of these tests must be identified and controls must be in place to ensure that they have been destroyed.

Personalization

Personalization of cards may be carried out by the card manufacturer, by a specialist company, or by the user or scheme operator. Personalization involves adding the individual cardholder data; it is not required for an anonymous telephone card, for example.

Data transmission

We assume that the raw data are held on file by the end-user company. The first question to ask is whether the data to be stored on the card are sensitive: In this case it will probably be protected by some form of encryption or access control on the original system.

When the data are transmitted to the personalization company or department, the same level of security must be maintained. Some form of MAC or integrity check is probably desirable in all cases because it may not be possible to correct any wrong data written to the card. Where the data pass outside the company, a simple encrypting modem or a secure tape transmission service should normally be used.

Fixed and derived data

Some of the data may be added or calculated by the personalization company. This may include fixed data pertaining to the card issuer (the issuer profile). For multifunction cards, the data to be loaded in one phase of personalization may depend on data read from the card as a result of a previous phase.

Derived keys or certificates (see Chapter 5) should be generated inside the personalization device or a host security module attached to it. Similarly, where asymmetric keys are used and the private key is to be stored on the card, the keys should be generated by the personalization device or on the card itself. In this way, only the public key need be transmitted back to the scheme operator or host computer.

Testing

Once the data are loaded, the card is again tested. The tests may include functional tests (it is not always possible to test all of the card's functions before personalization), visual tests, and—very importantly—*synchronicity* tests, which ensure that the data written to the chip correspond to any data written to the magnetic stripe or printed on the card. Card personalization machines often have a carousel with several stations, which can allow cards to get out of sequence unless this check is carried out. Where cards will be issued by mail, the mailers are usually printed at the same time as the card is personalized, and again synchronicity must be checked after the cards are mounted in the mailers.

It is critical to track all failures at this stage of testing, as cards rejected because of a visual defect or synchronicity check may be completely functional. Development departments often seek to use such cards for testing, but this should be avoided at all costs!

As with the previous stages, a personalization number and unique reference is added, and once the personalization process is complete, the personalization lock is blown.

The data loaded onto the card (not including private keys) should usually be stored securely for a short time after the card has been personalized. This enables audit and recovery in case of problems. In some cases, data may need to be transmitted back to the host system. In any event, the card number and card and batch status information should be returned for control purposes.

Data protection

Data-protection requirements must be taken into account during card personalization. We must first decide whether any of the data are regarded as personal, and if so whether the cardholder's permission has been obtained to store it on a computer system. According to the relevant national legislation, it may be necessary to provide a way of giving cardholders access to personal data stored on the card or to protect certain fields according to the wishes of the cardholder.

Electrostatic discharge and interference

Throughout the manufacture and personalization stages, and indeed throughout the life of the card, we must remember that a smart card contains a semiconductor device. It is therefore susceptible to problems such as electrostatic discharge (ESD) and interference from other devices.

Equipment that was designed to handle traditional magnetic-stripe cards may not prevent, and may even cause, such problems, and its suitability must be reassessed.

Issuance

Cards may be issued to customers by a variety of methods, such as over a counter, through dispensing machines, and by mail.

In all cases, a control should be in place to register the card's change in status. Sometimes a bar code or other mechanism is used to identify the card at the time of issuance; it need not necessarily be performed through the chip itself. In the case of postal dispatch, we know that the card has been sent, but not that it has been received; it is sometimes valuable to recognize these two different states.

Organizations that issue cards will already be familiar with the common forms of postal interception (e.g., accommodation addresses, false notification of changes of address) and the methods that can be used to defeat them. Two very common methods are to have a time gap between the card being issued and the beginning of its validity (during which time the issuer checks that the card has been received), or to require a separate action that includes an identity check (a telephone call, a visit to a branch, or use in an on-line terminal) to activate the card.

Smart cards can sometimes be damaged in postal distribution systems: In particular, high-speed sorting machines often cause mail to turn a sharp bend when it hits the right pocket. Mailers should be designed to provide some form of protection, and where possible the card should be mounted so that the chip faces away from the front of the envelope to minimize damage from this source.

Loading/validation

Many types of cards (in particular electronic purses and stored-value cards) must be loaded before they can be used. This may be done at an ATM or other on-line terminal, at an off-line *card validation station*, or by telephone.

This is a standard double-entry bookkeeping procedure: A credit entry on the card account is matched by a debit to cash or to the account from which the money was drawn. As we will see, some purse schemes hold a *shadow account* on the central system, which should match the account held on the card.

The procedures for loading a stored-value card or purse usually involve higher security than when the card is used. Cards tend to be loaded with one large amount, which is then used in many transactions. Often a PIN or other password will be required for loading, even when the card may be used freely without a PIN. For example, customers with reloadable telephone cards are likely to be asked for a PIN when they press the special button that transfers a fixed amount of money from the account specified on the card or customer record to the card.

Use

As the card is used, we need to address the issues of logging and audit, card and cardholder authentication, and recovery from errors.

Logging

To maintain the complete record of the card's status, the first use of a card is often regarded as significant; this tells us that a transaction log for this card must now be opened and may give an opportunity to carry out extra checks to ensure that the card is being used by the person to whom it was issued.

When a card is used only as an identifier, it is unlikely that any data will change, and in these cases the use of the card need not be logged. Actions performed with the card may be logged for other reasons: For example, in an access control application, we may record the door being opened or a person arriving at work.

Where the operation results in data being changed, however, we usually record the change. If we are dealing with a transaction that will be logged in full somewhere else in the system, then it may be enough to increment a transaction counter within the card. This allows each transaction to be identified and provides a check that the record of transactions is complete. If the central transaction record does not exist or may be incomplete, the full transaction details should probably be stored on the card. Sometimes only the last five or ten transactions are stored in a rolling buffer, but for maximum security and audit, or if it may be necessary to deal with customer inquiries, a complete record should be held on the card. This may increase the memory requirements within the chip considerably.

Card and cardholder authentication

Whenever a card is used, the system or terminal should invoke whatever card authentication checks have been used—examples of these will be discussed in Chapters 12 to 18. This allows the card to identify itself positively to the terminal, and it may also be necessary to have the card authenticate the terminal (to prevent bogus terminals being used to extract data from cards). If there is a cardholder verification method (a biometric, password, or PIN check) on the card, this may be required for some operations but not for others. Thus, the decision to invoke this check can only be taken after the application has been selected, and it is usually part of the application logic.

Error recovery

The design of the transaction itself must take into account the need to recover from errors during the transaction, particularly where the transaction is not completed or where the card loses power during the transaction. This is common for contactless cards, but can also happen for a contact card if the cardholder is impatient and withdraws the card too quickly. In these cases, we usually want to cancel the transaction and return the card to its previous status, but we cannot do this if data have already been changed.

As for network errors, we can either use a transaction log or the two-stage commit technique. The first method is adequate for simpler applications, but for more complex systems, and particularly for functions that affect more than one application or data file on the card, the second method is considerably more secure and makes recovery easier.

Lost, stolen, and misused cards

Issues

When a card is lost or stolen, the outcome will depend greatly on the type of application loaded on the card. First we must be able to detect a lost or stolen card. In most cases we can prevent a thief from using a stolen card, and sometimes it will be possible to restore any value stored on the card to the genuine customer. In any event, we need to be able to reissue a card to the customer.

For anonymous cards (such as disposable telephone cards) none of these is likely to be possible—when a card is lost, the finder gains its value.

First it may be necessary to verify that the card really has been lost: There are situations in which a user may benefit from reporting a card lost or stolen (for example, to deny responsibility for subsequent transactions on the card). A police report number or other proof may be required. The genuine user may also misuse the card, breach the conditions of the card-holding contract, or have the card withdrawn because of resignation. In all these cases we need to invalidate the card and prevent its use.

The situation becomes more complicated in the case of multifunction cards: The card may still be valid for other applications when one is blocked. Even if the card issuer contract is broken, the card may still contain value from another application issuer. So we need to be able to distinguish between actions that block the entire card and those that block only one application.

Detection

Ideally, the cardholder notices as soon as the card is lost or stolen and reports it to the scheme operator. The card is then put on a hotlist, and any on-line transaction will immediately be detected and stopped. In many systems, the hotlist is distributed to all terminals, immediately or at regular intervals; the terminal can then take action to block the card or can perform an on-line transaction to allow the scheme operator to take appropriate action.

Since cardholders will not always report the loss, and to guard against misuse, the scheme operator must also analyze incoming transactions to spot unusual patterns. For example, access-control systems look for con-secutive uses too close together, while bank card systems look for a rapid succession of transactions or a change in customer behavior. Some credit card issuers employ neural networks to spot high-risk transactions and enforce a further check.

Block and unblock

If a stolen or misused card is detected, the terminal or host computer system can set flags that will block either the application or the entire card. It is fairly easy to reset a blocked application when the problem is sorted out (this might therefore be used in the case where a customer exceeds a credit limit); unblocking the card usually requires a further level of security because it affects all the applications on the card.

Financial cards are often blocked by sending the card a sequence of wrong PINs, until the card refuses to accept any more attempts. The corresponding method of unblocking is for the card issuer to give the cardholder a PIN unblocking key (PUK). This key (which has previously been stored in the card) allows the PIN-try counter to be reset to zero only once; once a PUK has been used, a new one must be generated and stored in the card.

When a card or application is blocked, the card status must be changed accordingly, and usually an attempt will be made to recover the card or rectify the problem. Blocking procedures should be designed in such a way that they can be carried out at any terminal. Unblocking must be a more tightly controlled operation and is often only possible at specific terminals or after entry of a further security code by an employee of the scheme operator.

Reissue

Once a card has been reported lost or stolen (whether or not it has been blocked or recovered), there must be a procedure for issuing a replacement card. In some cases, this will be identical to the procedure for initial issue, but in other cases we must make provision for restoring the value on the card to the customer or application issuer.

For schemes with full *shadow accounting* (see Chapter 13), this will be an easy task, although often we must allow a sufficient period to elapse for all outstanding transactions to have been registered. In the absence of shadow accounts, it is virtually impossible.

Application loading and unloading

For multiapplication cards, we must have procedures for loading new applications after issue, updating applications, and unloading them to make space for new applications.

Ideally, we would like to replicate the security of the original personalization environment when loading new applications onto the card. But this is not usually possible, except perhaps for company or other closed user group cards. The whole point of multiapplication cards is the convenience they offer—customers want to be able to carry out these operations with minimum disruption to their routine. Even going into the bank branch rather than using an ATM, or going to the customer-service desk in a retail store, may be enough to prevent them from taking the new application.

At the other extreme, card issuers are unlikely to be keen on customers loading new applications onto their cards at their home PCs—we must always assume that the contacts on the card reader can be "bugged" and so any programs or authentication keys could be copied. The Multos system is able optionally to transmit the whole application in encrypted form and only decrypt it following secure authentication (see Figure 11.3).

Card issuers will in most cases insist on any new applications being loaded in a secure environment, such as an ATM, customer-service desk, or specialized card-loading center. The application load procedure is determined by the multiapplication OS in use; it will normally involve an authentication procedure to ensure that the application has been approved by the card issuer for loading onto that card; this may take place on-line or, as in the Multos case, the card may check a certificate previously generated.

The procedure for updating and deleting applications follows generally the same principles, but the levels of control may be slightly lower in some cases.

All the procedures proposed to date have been centered on the need to maintain card-issuer control. No one has yet created generic procedures and systems suitable for the consumer-owned "white card," to which we will come back in Chapter 18.

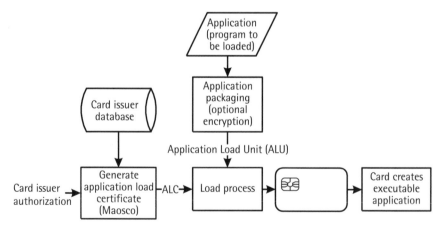

Figure 11.3 Application load sequence for Multos card.

End of life

Expiration

Cards should be given an expiration date. It may be possible to extend this date on-line or to ignore it in use, but the existence of a date is a final form of protection against cards that have defeated all forms of misuse detection. It also provides a natural route for upgrading cards and applications as technology and services move on. Bank cards are usually given expiration dates up to three years from issue. The cards themselves are usually expected to last longer than this, but some cards will already be showing signs of unreliability by this stage. In addition, it is difficult to extend the card's validity where the expiration date is printed or embossed on the card.

With multiapplication cards, we must ensure that no application has a longer life than either the card itself or any master application. Ideally, they should all terminate on the same date. This is not always easy to organize.

Dispose or recover

Even an expired card may contain secret keys or algorithms, which may be of use to a hacker. A stack of expired cards allows a test to be repeated many times or carried out in parallel, whereas few hackers will risk experiments on their own or their friends' valid cards! It is therefore wise for a scheme operator to attempt to recover or have destroyed any cards that expire or fail.

For a closed user group such as a staff card, this should not be difficult. For other user groups, it may be necessary to arrange an incentive scheme, to issue cards only on an exchange basis, or to have a special transaction that validates the new card in return for capturing the old one. If none of these is possible, serious consideration should be given to the life of any keys used in the card; it may be necessary to update master keys and private keys at least as fast as the card expiration cycle.

Recycle

Card materials, particularly the PVC used in bank cards, are environmentally damaging, but recovered cards are often still fully functional. In some schemes it will be possible to reuse cards with very little effort, but in the majority of cases the card must be destroyed. Most manufacturers now have schemes for recycling a high proportion of card materials.

Part III
Applications

12

Telephony and Broadcasting Applications

In the next few chapters we consider some of the most important applications in which smart cards are used. For each application, we consider the security requirements and major issues, and how smart cards are or may be used to address these requirements. In this chapter we look at the smart card applications that account for the majority of cards in use today: those in telephony and broadcasting.

Smart cards are used in these industries for three main applications: as telephone cards for public (fixed) telephones, as subscriber identity modules (SIMs) in mobiles, and as conditional access (CA) modules in cable and satellite-broadcast networks.

Fixed telephones

Public telephone requirements

Public telephones using cash are expensive to build (they must be very robust to protect the cash from theft), expensive to operate (because of the need for cash collection), and unreliable (cash mechanisms become full, jammed, or vandalized). For many years now public telephone operators have exploited various forms of cards or tokens to overcome these problems.

Card telephones should have a minimum number of moving parts and must be able to operate reliably in a wide range of environmental conditions. The cards themselves must be easy to handle for all types of users, and they must be more costly to counterfeit than the maximum value on the card (typically $20 or so). Although various forms of magnetic and optical cards have been used with success over the years, most telephone operators have now opted for smart cards as the most effective card form (see Figure 12.1).

Although these have not completely eliminated fraud (many types of fraud exploit the network rather than the payment method), fraud in telephone networks is now at a historical low.

Telephone card standards

The so-called first generation of telephone cards was made up of simple memory cards with an issuer identification area, logic that prevents the value on the card from being increased, and usually a relatively small maximum count (100–200 bits). The second generation (based on the French and German T2G specifications) had much higher maximum counts (up

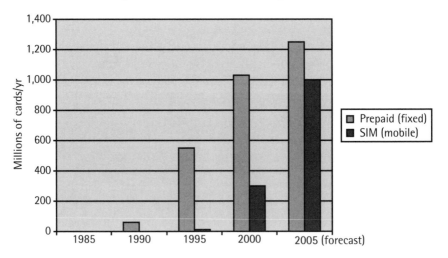

Figure 12.1 Market for smart telephone cards in Europe. (*Source:* Author from compilation of industry statistics.)

to 20,000 bits), card authentication mechanisms that protected against counterfeiting and emulation, and logic to protect the card in case it is removed during a transaction.

The current, third-generation cards are based on the Infineon Eurochip and other companies' variants. These have the same facilities as the second-generation card, but also include a dynamic "challenge and response" authentication method (see Chapter 5), which makes counterfeiting virtually impossible, along with a user memory area allowing number storage.

The particular benefit of the third-generation cards (apart from their higher security) is that each telephone company issuing cards can have its own set of private keys, but telephones can store the public keys of several different issuers, thus allowing cards to be used in different countries or telephone operators' systems.

Issues

Telephone cards are often sold in small denominations ($2 or even less). Although the cost of cards is falling steadily as volumes increase, the cost of the card is still significant in relation to the value stored. So it is in the interest of telecommunication operators to maximize the value on the card and to add to the functions available on the card.

Both cards and telephones must have extremely high reliability, even though they are often misused. To avoid accidental misuse, the cards must be very easy and instinctive to use; it must, for example, be very clear when a card is correctly in position. Most telephone cards now incorporate a special tactile identifier (see Figure 12.2) that indicates the card orientation. The cards must also be able to handle such problems as being withdrawn at any time during a call; the status of the card at all stages in the transaction must therefore be recorded.

To minimize the overhead on the lines, card telephones must normally operate off-line during a call and only exchange control information with a host computer on a periodic basis (e.g., once a day). Some modern systems are able to remove this limitation, at least partially, by using digital lines and distributed control systems. Nevertheless, the overhead involved in managing each call must be kept to a minimum. Individual details of cards used are not normally regarded as essential information for billing or operations.

When individual telephone cards are stolen, there is little the operator can do: The value is lost to the owner and gained by the thief. When large

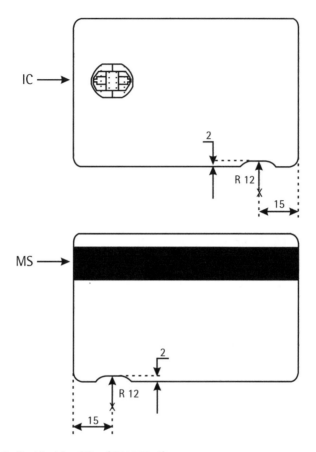

Figure 12.2 Tactile identifier (EN1332-2).

quantities of cards are stolen, however, they can be blocked by hotlisting if the card numbers have been traced throughout the distribution chain.

Telephone card developments

Traditional telephone cards have been disposable. The newer cards, particularly those with onboard dynamic authentication, can also be securely reloaded. Options include cash, loading from a linked account or from another card, such as a credit or debit card. Reloadability extends the life of the card but increases the unit cost; many telecommunications companies feel that the cost of putting in place systems for reloading cards is not justified. Reloading may also lead to lower long-term reliability and even to possible security breaches (most likely in the surrounding systems).

Bank cards in France (using the chip-based national debit application, which is described in Chapter 14) can already be used in France Télécom public telephones. As chip-based bank cards become more common, this should be feasible in other countries. There are two options:

- Add a telephony application to the bank card (as in the Belgian Proton card), thus value can then be transferred from the main card application to the telephony application.

- Have the telephones accept bank cards in their normal debit or credit mode. There may be some certification problems with this approach, since the banks' EMV standards currently conflict with the mode of operation used by telephones.

Smart telephones, which have a screen and small keypad, increase the range of services which can be offered through the telephone. These exist both for domestic use and as public kiosks (where fax and e-mail are the most commonly used facilities). Most smart phones include chip card acceptors, although there is as yet little standardization of the use of the card. The Pay Terminals and Systems project within the European Telecommunications Standards Institute (ETSI) is seeking to extend the EN726 standard to cover these applications.

There is a gradual convergence of fixed and mobile services in progress. This raises the prospect of customers being able to insert a personal SIM card into a fixed telephone or smart phone anywhere in the world, and being able to access the facilities of the telephone using the customer's own numbers and applications stored in the SIM. The fixed public telephone becomes another device accessible to the mobile user.

As we progress into the domain of authenticated transactions and encrypted communications, such facilities may become standard features of telephone cards and systems.

Mobile telephones

Cellular telephony was introduced to most European countries at the beginning of the 1980s and rapidly proved very popular. The early analog systems were different from country to country, and were prone both to eavesdropping on traffic and to calls fraudulently charged to another person's account, which resulted in significant losses for both the telephone operators and their customers.

When the design for a Europe-wide mobile telephone system was being discussed and agreed, therefore, security was a high priority. The global system for mobile telephony (GSM) is now the most widely used standard, although a different standard, PCS, is used in North America and Japan. The success of GSM has exceeded all expectations; penetration in major European markets is consistently over 25%, and even in South Korea more than 60% of the population has a mobile telephone.

Further background on the GSM system can be found in [1–3].

Subscriber identity module

GSM security aims to:

♦ Authenticate the user (rather than the telephone);

♦ Protect the integrity and reliability of calls;

♦ Protect the confidentiality of calls and call-related data.

It does this by using a smart card known as a subscriber identity module (SIM) in every GSM telephone. SIMs are either supplied as modules or complete ISO standard cards. A subscriber may have more than one SIM (for different networks, for example) or may use one SIM in different telephones (where the form factor of the card permits).

Most current GSM telephones use a Phase 2 or Phase 2+ SIM. Phase 2 has 8 KB of user memory for user data and telephone numbers. GSM defines its own operating system, which protects the data in the card so that it can only be accessed by authorized applications. Although the structure would permit applications on the card to be issued by different application issuers, no current schemes use this facility. The SIM specification (GSM 11.11) supports PIN checking, application and card blocking, and unblocking by means of a PUK (as described in Chapter 10).

SIM Toolkit

Phase 2+ adds a facility called the SIM Toolkit (STK) as well as extending the card memory to a minimum 16 KB. STK provides a standard interface between applications running in the SIM and the telephone's display, keypad, etc. It allows data and applications to be downloaded to the SIM; these can then be invoked by the user or by a command sent over the air, adding facilities to the telephone. Display formats can be tailored, and do not have to conform to the standard short message display. STK is ideal for applications such as access to a bank account, or for a menu of information services; frequently used data, logos, etc. can be stored in the SIM and do not have to be downloaded at each use.

Wireless Application Protocol

The next stage in the evolution of mobile services is the Wireless Application Protocol (WAP)—see [4, 5]. WAP allows telephones to display a subset of the Hypertext Markup Language (HTML) used for Internet pages. The Wireless Markup Language (WML) can only display text and small graphics, and navigation is also very limited. Nevertheless, it represents the possibility for mobile users of accessing a very wide range of information anywhere on the Internet.

Most current mobile networks use the short message service (SMS) for sending data; this has a maximum speed of 9,600 bps and limits services such as WAP. With the introduction of general packet radio switching (GPRS), data rates will increase by a factor of more than 10, and the packet nature of the service means that connection times will no longer be a limitation either. GPRS services and telephones are due to become available early in 2001.

Many banks and service companies have seen WAP as a tool for introducing mobile transactional services. For this purpose, however, WAP V1.1 suffers from security weaknesses and is only seen by the WAP Forum (the body responsible for the WAP standards) as an interim solution. The Wireless Transport Layer Security mechanism (which parallels the Internet's TLS) allows the WAP gateway to authenticate the mobile and vice versa, but this authentication is not carried across the WAP gateway to the service provider—the authentication is outside the service provider's domain (see Figure 12.3). The secret keys exchanged at the beginning a session may have quite a long life (days), and are not necessarily stored securely. And it seems that on some WAP telephones the user can turn authentication off, without the service provider being aware.

WAP V1.2 provides for a wireless identity module (WIM), which is a security token assumed normally to be carried as an application in the user's SIM or another smart card. WIM includes authentication of the user (normally by a PIN) and allows secret keys to be stored in a tamper-proof environment. However, this does not overcome the nonoverlapping domain problem described above.

Most service operators are therefore opting for authentication using the SIM Toolkit (which allows a direct connection between service provider and telephone—see Figure 12.4) but possibly making use of the WIM

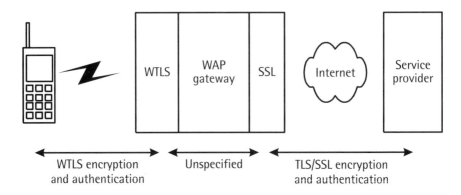

Figure 12.3 Nonoverlapping authentication domains in WAP.

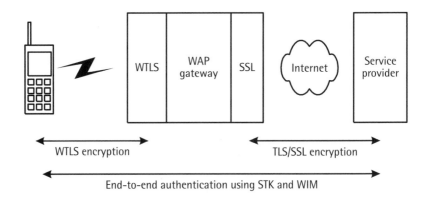

Figure 12.4 End-to-end authentication under service provider control.

application for secure key storage [6]. This is the format adopted by the Radicchio initiative, which aims to provide a standard security platform for mobile commerce.

Version 1.3 of WAP will include a full wireless public key infrastructure, with keys and certificates stored in a certification authority system rather than in the mobile. This will address most of these issues, but is not likely to appear before late 2001.

Universal mobile telephone service

The third generation of mobile telephones (sometimes known as 3G) will use the universal mobile telephone service (UMTS) standard, which is now close to final agreement. UMTS will replace both the PCS and GSM systems, combining the best features of each. The SIM has been one of the defining factors in the success of GSM, and UMTS will use a SIM.

Prepay

Many mobile telephone users do not have a personal account with a service provider; instead they prepay for services [7]. The network provider still maintains an account for the SIM in question, and this account may have time and call value.

There are several options for adding "top-up" value to a prepay account.

- Many network operators sell prepay tokens in the form of "scratch cards." The value is added to the account by keying the number on the card into an interactive voice response (IVR) system using the mobile or SIM in question.

- Value may be transferred from a credit or debit card previously linked to the account or at an ATM (this removes the anonymity which many people value in a prepay telephone).

- Some operators have set up point-of-sale networks, allowing customers to pay cash (or using another form of payment); the telephone number or account is identified using a magnetic-stripe "network card" and the transaction takes place on-line to the network operator's systems.

All of these methods present security issues: The scratch cards represent a large amount of unsecured value, and are almost impossible to audit in use. Because owners of prepay telephones are anonymous, a thief can use a stolen credit or debit card number to perform top-up operations with relatively little risk of being caught. (This resulted in major losses for one U.K. network operator in 1998.) On-line activation is the most secure, but even here there are issues surrounding incomplete transactions and refunds. The operators therefore perform transactions on a buffer system, transferring the value only after a certain time period (say, 10 minutes).

Two-slot telephones

As most mobile phone users carry their phones at all times, and there are now more mobile phones than bank cards, the mobile is being seen as a preferred way of carrying out transactions.

Certainly the mobile can be used for selecting goods, for entering a PIN or other security code, and for securing the data transmission. However, the issue which has not yet been resolved is that of the ownership of the payment medium. Up to now, banks have assumed that they were the sole issuers of payment cards—their role is protected by law in many countries. However, if the mobile phone is to be used for payment, then where does the payment card fit in?

Do we have a bank card application resident on the telecommunication company's SIM card? Or a SIM application on a bank-issued card? Can we make a link from the SIM to a bank server, carrying the customer's identity?

Similar questions arise if the mobile is to be used for authorizing purchases, giving trading instructions, or signing e-mails. In each case the application needs to have access to a second card application, issued by someone other than the network operator.

Several telephone manufacturers have responded by fitting telephones with two chip card reading slots: one for the SIM and the other for a bank card, employer's card, or other token (see Figure 12.5). A number of pilot schemes use this facility, and many commentators have forecast that all telephones will have two slots within a few years.

Cable and satellite television

Requirements

The growth of subscription television has provided another opportunity for fraud and subscription evasion, this time at the expense of the broadcasters and service providers. Although much smaller than telephone fraud, some companies estimate their losses at up to $20 million a year.

Satellite and cable television companies seek to obtain a proportion of their revenues from subscriptions, particularly for premium services, and *pay-per-view* income (although advertising remains the main source of revenue for most companies). Subscription services are transmitted in an encrypted form, and the decryption is carried out in a *set-top box*, which is fitted with a smart card (see Figure 12.6). The cards are sold on a subscription basis and may also contain credits for pay-per-view watching.

The smart card also provides the means to control other aspects of viewing, such as programs suitable for children. Some industry sources

Figure 12.5   Dual-slot GSM telephone. (Photograph courtesy of Motorola Ltd.)

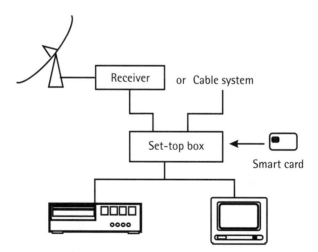

Figure 12.6   Television decryption.

also believe that it will allow selective downloading of advertising or messaging or even emergency messages broadcast to specific areas or groups. Although satellite signals reach all countries, governments may control the sale of cards for specific services thought to be undesirable.

Weaknesses and responses

The Videocrypt encryption method and keys used in analog satellite television have been repeatedly cracked; software containing the algorithms and codes has been published on the Internet, and pirate cards are sold openly in several countries, notably in Germany. A North American pirate card ring raided in mid-1996 was estimated to have produced over 30,000 cards. Figures of $20 million per year quoted for lost revenue to the networks are probably exaggerated, as most of those who bought the pirate cards would not have paid the full price for the service. The cost of replacing the smart cards was much less than the cost would have been if the whole set-top box had needed replacement.

Nevertheless, this is the most serious threat to the credibility of smart cards as an aid to security, and it is important to know what lessons have been learned.

Much of the problem can be traced to an underlying weakness in the encryption method. Modern encryption relies on digital techniques and is applied to digital data. When applying it to analog television signals, the original designers simply divided the signal into finite slices. Each television line is divided into two parts, and the two parts are rotated. Every few seconds a new key is transmitted in an encrypted form in the vertical blanking interval (the gap between pictures usually used by teletext and overlay signals); this is decoded using the master key held in a smart card in the set-top box.

There is therefore a strong correlation between the encrypted signals and the original; it is even possible, with a high-speed signal processor, to match elements of successive lines so that the signal can be decrypted without use of the algorithm at all! For the hacker, however, the availability of large quantities of matching plaintext and ciphertext reduces the key search time to very low levels once the algorithm is known. The second level of encryption involved in deriving the master keys can also be simplified by building a unit that appears to the card like the set-top box, but which is directly controlled from a PC.

These cards were replaced in 1994 with cards using two-way authentication, a much stronger algorithm, and longer keys; since then the incidence of hacking has been reduced, but for most analog systems there are still pirate cards available.

The move to digital

Today, analog television broadcasts are rapidly being replaced by digital transmission, where the full range of modern encryption techniques can be used. The universal standard for digital broadcasting is the digital video broadcast (DVB) standard [8, 9], which allows digital enveloping of an MPEG-2 (Motion Picture Expert Group level 2) signal. DVB also includes provision for a conditional access (CA) system at the transport layer, although there are several proprietary implementations of this CA system, from companies such as NTL, Philips (Cryptoworks), Irdeto, Telenor (Conax), and France Télécom (Viaccess).

All the schemes use chip cards to store long-term keys, which can be updated or canceled over the air, and which allow decoding of the short-term keys actually used to encrypt the data. The mechanism for updating the keys and encrypting the data varies from scheme to scheme.

In order to allow set-top box manufacturers to design receivers that will operate with all these CA schemes, DVB defines a common interface (CI), in the form of a PCMCIA receptacle. There are CI modules for most of the common CA schemes, and these in turn accept the subscribers' chip cards.

The DVB standard allows data to be carried in several different ways alongside the television signals in a DVB transport stream. This allows channels to carry not only information alongside their services, but also advertising, software, and other data.

Completely independent data streams can also be carried on the same signal: these may include financial market information, distance learning, or even interactive services—in the last case a return channel is needed, usually through a modem attached to the PC or set-top box. Full Internet services can even be carried through DVB channels; the downstream data rate (reception of data) may be up to 34 Mbps, directly to a single user's home or office.

In these cases, there are further security issues to address: Satellite signals can be intercepted by anyone within the beam, and if the IP address of the target receiver is known, the transmission can readily be extracted from the television signal. So a further level of encryption is needed for most IP traffic carried on these channels. Because of the problems of key distribution and management, a public key system is preferred, and there are now public key algorithms which can encrypt and decrypt at the speeds required; otherwise a public key method is used to exchange a symmetric encryption key.

In the longer term, a public key infrastructure needs to be defined for DVB data applications.

References

[1]   Redl, S., M. Weber, and M. Oliphant, *GSM and Personal Communications Handbook*, Norwood, MA: Artech House, 1998.

[2]   Heine, G., *GSM Networks: Protocols, Terminology, and Implementation*, Norwood, MA: Artech House, 1999.

[3]   Schiller, J., *Mobile Communications*, Reading, MA: Addison-Wesley, 2000.

[4]   van der Heijden, M., and M. Taylor, *Understanding WAP: Wireless Applications, Devices and Services*, Norwood, MA: Artech House, 2000.

[5]   Arehart, C., et al., *Professional WAP*, Wrox Press, 2000.

[6]   Birch, D., "Telco as Trustco—PKI Business Opportunities for Telecoms Operators," *Proc. Tele-Commerce*, Geneva, May 2000.

[7]   Hendry, M., "Prepaid Mobile Phones—Electronic Topups Taking Off," *Cards Now Asia*, October 2000.

[8]   de Bruin, R., and J. Smits, *Digital Video Broadcasting: Technology, Standards, and Regulations*, Norwood, MA: Artech House, 1999.

[9]   Reimers, U. (ed.), *Digital Video Broadcasting*, Berlin: Springer-Verlag, 1996.

13

Computer Networks and E-commerce

Anyone operating or responsible for a computer network today (and that means at least one person in almost every business or organization) is conscious of the need for security. But many people are unsure what the risks are for them, or how to overcome those risks.

The massive publicity given to virus scares, for example, might make companies concentrate on installing virus protection and e-mail proxies. Or, the fear of hackers might lead companies to install firewalls, and then feel protected. A recent consultancy assignment involved a company that wanted to protect its Web site—essentially data that the company wanted outsiders to view. They had proposed using firewalls, but on closer inspection it was clear that the issue was not "Who can view this information?" but "Who can alter this information?" Since only a few people in the marketing department had the authority to alter the information, the primary solution lay in an authentication scheme rather than in the firewall functionality (the solution was in fact implemented using a firewall because of other considerations).

Network security

The first problem with IT security is persuading managers that there is a potential problem—often senior managers are the worst offenders! The reports produced each year by the National Computing Center in the United Kingdom [1], by the Computer Security Institute in the United States [2], and similar organizations are a good starting point—they usually show that well over half of all organizations have lost money as a result of security incidents.

Most IT security incidents, and nearly all those resulting in high cost to the organization, involve breaches from within rather than external hacking. However the problem is posed, most network security issues involve authentication to some degree. The British Standard/ISO Code of Practice for IT Security Management [3] places considerable emphasis on the control of user access and privileges; these can only be enforced if the network administrator or data owner can rely on the initial user identification. It is increasingly common for data ownership to be very widely distributed through an organization, and this is where an authentication scheme will prove its worth.

Computer system access

Most computers are accessed by a combination of a user identification (user ID) and password. Enforcing good password discipline is notoriously difficult; the same controls that make it difficult for outsiders to guess passwords also make the passwords difficult to remember. Even with good discipline, valid passwords can be obtained by looking over people's shoulders, by monitoring data traffic, or in some cases by hacking into system files.

A much more secure form of control involves the use of an intelligent token in conjunction with a password. Now that smart card readers are becoming easily available for PCs, they are one of the easiest forms of tokens to implement. The password (or a PIN number) is checked by the card and need not be stored anywhere on the system. When the card is inserted, the system asks for the user ID; this may be given by the card or by the user. The system then authenticates the card (as described in Chapter 5), and the card authenticates the user. The use of random or time-dependent functions is important to avoid replay attacks.

The card may also contain a user profile, including preferences and membership of groups (where access to specific system functions is specified by groups). The access rights themselves should be held within the system and associated with the data files or applications in question.

When the card is removed, the user should be logged off.

There are several proprietary authentication schemes on the market, and these are regularly reviewed in computer security magazines (for example, [4]). A smart card–based sign-on mechanism of this type has been built into the NT operating system since 1998, and now forms a standard option in Windows 2000.

Where a card is used, it can store a range of passwords and authentication mechanisms. The user authenticates him or herself to the card once, and then has no need to remember the dozens of passwords which he or she may need to cross firewalls, access different Web sites, authorize payments, and so forth. The original authentication must be completely reliable in this case—a fingerprint or other biometric should be used rather than a password.

Confidentiality of data and programs

File encryption is often used for protecting stored data. Again, there are many commercial packages available for this, most of which use a password to access the encrypted files.

Systems for protecting stored data must always strike a balance between confidentiality and ease of use. If a system is designed to protect against every possible attack, the smallest malfunction can render the data unreadable. Storage of keys is a particular problem and is an area where smart cards can be of assistance. They also help to enforce certain specific types of access control, notably where individuals are held responsible for authorizing access to data. In these cases, the keys are held on a smart card, which is kept by the person responsible. When designing such a scheme, it is important to set out the security objectives clearly and to acknowledge, where tradeoffs must be made, what decision has been taken.

Smart cards are also used for storage and loading of keys into encrypting modems and other hardware devices used for encrypting data during transmission from one system to another. The procedures described by the manufacturers of these devices for setting up and maintaining these systems must be carefully followed if their security is to be maintained.

Internet browsing and e-mail

Internet protocols

The protocol used for passing messages between computer systems on the Internet is called Transmission Control Protocol/Internet Protocol (TCP/IP). TCP is concerned with message handling and IP with addressing. TCP/IP was designed to allow messages to be routed through the network using any available connection; it thus offers a high level of resilience and alternative routing, but there is no protection against any of the intermediate computers reading, altering, or destroying the message.

Current Internet addresses (where www.abc.com is translated into 123.123.123.123) are running out, in the same way that telephone numbers run out. The next version of the Internet protocol (IP version 6) will not only extend the addressing space, but will also include a range of security features, including the ability to "hide" ultimate destination addresses (so that sites and pages can only be reached by following designated links). IP v6 will also support the IPSec authentication protocol, which provides full end-to-end encryption and authentication. Sites and users using IP v6 can communicate as though they were on private networks. The problem with IP v6 is that it must be supported not only by the servers and browsers, but also by all the intermediate routers and gateways. The whole Internet infrastructure must be upgraded, or a sufficient amount of it to provide a path between users wanting to use IP v6, so it will take some years before open connections can rely on its availability.

Secure sockets layer

Although the possibility of eavesdropping, or "packet sniffing," is often quoted as a source of insecurity on the Internet, a protocol that overcomes this has been available on all common browsers since the mid-1990s.

All of these browsers offer Secure Sockets Layer (SSL) or the equivalent Transport Layer Security (TLS) protocols. SSL and TLS encrypt data between the browser and the host, providing both confidentiality and integrity checking. They also provide a level of authentication: browser and host authenticate each other, and the browser will only work with a server certificate issued by a CA it recognizes.

However, such server certificates are relatively easily obtainable and make no statement about the probity or financial standing of the organization to which they were issued. Few users even check the identity stored in the certificate (which can be done by clicking on the padlock which closes when a secure session is in force). And the SSL protocol does not give servers any information about the identity of the user accessing the information.

Public key infrastructures

Provision and checking of this extra information about the identity of servers and users is the domain of public key infrastructure (PKI) systems. PKIs were mentioned in Chapter 5, and network administrators can buy off-the-shelf packages from major-name players such as RSA Data Security, Entrust, IBM, and Baltimore, to set up and run their own PKIs. These are strongly recommended for major organizations with significant wide area networks.

In a fully open environment such as the Internet, private PKIs are not appropriate, and we must rely on Internet certification authorities recognized by browsers (a list of these is accessible within the security settings on the browser). Certificates issued by these CAs today do not contain sufficient information, nor are they backed up by sufficient warranties, for us to rely on them for serious commercial transactions. For a discussion of the legal issues, see [5].

Organizations such as Business-to-Business Exchanges wanting to secure groups of users, or specific types of transactions, must therefore issue certificates to their users under a PKI structure.

Most PKI packages include the option of using smart cards for carrying certificates, which gives them a much higher degree of portability. Users must be aware, however, that the X.509 v3 standard that governs the format of certificates contains a wide range of optional fields; certificates issued by one system cannot necessarily be read by another system, so the degree of portability offered even by a smart card system may be limited.

Hacking

Hacking has become the most common source of disruption for many sites, particularly those with some government or public-issue interest. Organizations can protect against this to a large extent by using firewalls, which only allow certain types of traffic and users to enter their systems. But firewalls do not prevent eavesdropping (other systems gaining access to data transmitted) unless the data is encrypted. And few firewalls are able to prevent files from being received if they are "invited" in. These may include cookies (programs that store data about the user and transmit that data to a host when the host computer is next accessed) and other nastier snooping programs (which intercept data and store it for later transmission).

The community of hackers includes people with many different motivations, ranging from financial gain or industrial espionage to the enjoyment of technical skills and prowess. Any system with specific pretensions to security is a target for hackers who seek a challenge. Hackers do not usually have access to the massive computing resources required for brute force attacks on cryptosystems; they therefore seek easy ways in, particularly ways that involve learning passwords.

One of the specific advantages of a smart card–based system is that it allows challenge-and-response authentication instead of simple passwords. Unlike a password, the response is never the same, and this defeats many potential attacks. Access control can also be by a combination of a token (the smart card) and a password or biometrics: When challenged, the card asks for and checks the password before issuing the response. Neither stealing the token nor gaining access to the password on its own is sufficient to gain access to the system.

E-mail

Probably the most widespread application on the Internet is electronic mail. Much e-mail is not confidential, and those sending and receiving it can risk the small chance that someone else reads or even alters it. In certain cases, however, a higher level of privacy or integrity must be ensured. This applies most often to commercial data such as orders, stock lists, and communications with field staff, and to subjects that are covered by data protection or similar legal restrictions (such as communications between doctors about a patient).

There are several standards for increasing the privacy of e-mail. Most of these include authentication of the originator and recipient, encryption of the data, and integrity checking of the whole message. They typically make use of a public key system for authentication and transmission of a symmetric key, which is then used for encrypting the body of the message. The systems most commonly used are the Secure Multipurpose Internet Mail Extensions (S/MIME), Privacy Enhanced Mail (PEM), Riordan's Internet Privacy Enhanced Mail (RIPEM), and Pretty Good Privacy (PGP). Most e-mail software systems will support one of these, although different versions must sometimes be used within and outside the United States for export control reasons.

The private key for authentication may be stored most conveniently on a smart card; it is then relatively secure and can be used on any computer fitted with a smart card reader. Some packages will now support this option. The public key can, of course, be published in an e-mail message or on a web page.

E-commerce

Although astronomic volumes and growth percentages are often quoted for e-commerce, the significance of these numbers is often lost. It is true that the numbers are high: Figure 13.1 shows the forecast growth in total business carried out over the Internet. But, as this figure also shows, business-to-business (B2B) e-commerce volumes are 10 times as great as those for business-to-consumer (B2C).

The issues involved for these two types of e-commerce are very different.

Business-to-business

The majority of B2B e-commerce takes place under preexisting contracts, often exchanged in writing beforehand. The Internet is only slowly taking over from the private electronic data interchange (EDI) networks that have

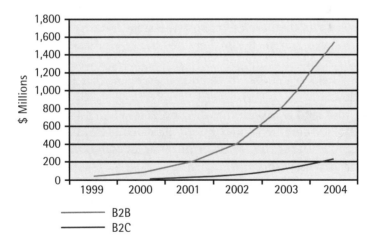

Figure 13.1 B2B and B2C European e-commerce volume growth forecast. (*Source:* Forrester Research.)

dominated this sector for many years; it offers the opportunity for smaller companies to participate without excessive cost, and allows suppliers to extend their sales reach without the high entry barriers associated with joining EDI networks.

B2B e-commerce is driven by purchasing systems and procedures. It rarely makes use of on-line payment—transactions are settled through an automated clearing house mechanism, using a standard invoicing and accounting procedure. However, positive authentication of the company and person placing the order is very important, and this is where card-based schemes such as purchasing cards are becoming popular.

Business-to-business exchanges (sometimes known as B2X e-commerce) can only expand if they have a means of easily accommodating new joiners, and as we identified earlier, this will often require the B2X exchange to put in place a public key infrastructure. In Europe, several Chambers of Commerce are participating in PKI schemes which will enable all their members to identify themselves to each other; some of these are issuing smart cards to their members for this purpose.

Business-to-consumer

The key issue in B2C e-commerce is to capture the sale in the fleeting moment when the seller has the user's attention. Many psychological factors are involved in this, not least of which is trust. Building an image of trust is difficult for a web site, and any one negative factor can put customers off and deter the sale.

Speed is another critical factor: the smaller the number of clicks or keystrokes required, the higher the chance of making the sale. But this requirement can conflict with that for security: If all the customer's details are held in a cookie, then anyone accessing that PC can carry out transactions.

The form and style of payment is one factor often quoted as inhibiting the growth of e-commerce, and indeed the statistics from the card schemes bear this out: Internet transactions are 50 times as likely to be disputed as face-to-face transactions.

A key problem, however, is simply that many people do not have the necessary form of payment: Most sites will only accept credit cards (or debit cards not requiring a PIN). In Europe, less than 15% of the population has such a card, including no one under 18. Many merchants are unable to accept cards because they do not have the necessary two years of audited accounts. And in only a few countries is there a mechanism for consumers to raise direct credits on-line other than through an on-line banking system.

Another problem relates to the size and nature of many Internet transactions. Many transactions are for very low values (less than $1), or are measured "by the click" or by the minute. This model will increase as on-line application service provision becomes more widespread.

Many Internet transactions are international, and every dispute is very costly for the banks and merchants involved. Merchants with a particularly bad record for disputes are likely to be fined or even have their agreements canceled. The international nature of the Internet means that companies have to be prepared to do business all around the world, under a wide variety of unfamiliar legal systems.

And there is often a need to identify the person carrying out the transaction: to prove that the address to which the goods will be delivered is that of the real customer, or that they are entitled to carry out this transaction (e.g., over 18 for gambling).

In the next chapter we will discuss the card schemes' approach to on-line card payments. But smart card solutions can also help to resolve many of these other issues.

Forms of payment

When making a payment for goods or services on the Internet, merchants may offer all or any of the following methods:

- *Account payment:* Customers set up an account, either by filling in a form on-line or by fax or mail. Following a credit check, they are allocated a credit limit and then given a user name and password for access to a secure area. Payments are made off-line, usually by direct debit.

- *Preregistered card payments:* As with an account payment, the card details are registered in advance, and these are called up whenever a transaction is made.

- *Open payments:* Card numbers and expiration dates can be sent to merchants openly, by e-mail or through a common graphical interface (cgi) page; this is definitely not recommended for any of the parties involved.

- *Secured link:* Most merchant sites now offer an SSL connection, which overcomes the eavesdropping risk. The merchant is responsible for securing the transaction thereafter, which means that the customer has to place some degree of trust in the merchant. (In order to clear the transaction, the merchant also has to have a banking relationship, and the bank will perform its own checks.)

+ *Trusted third party (TTP):* Several organizations have set themselves up as intermediaries; both merchants and customers must register with the TTP. Customers usually register a credit card or bank account number and continuous payment authority. The TTP allocates each party an account number, and is responsible for collecting funds from customers and crediting merchants. Only the account number is transmitted on-line, and this may be checked against a delivery address by the TTP. Variants on this model allow the customer to confirm delivery before releasing the funds (an *escrow* arrangement), or to make bill payments on-line (electronic bill presentment and payment—EBPP).

+ *Digital cash:* Again normally operated by a third party, digital cash consists of software certificates which are in effect virtual banknotes, with a fixed value. A customer must purchase these certificates in advance from the scheme operator. When a payment is to be made, the customer's software issues the appropriate number of certificates to the merchant. The merchant returns the certificates to the operator, which checks them and credits the merchant's account.

+ *Electronic purse:* These products will be discussed in more detail in Chapter 14. A payment can, however, be made directly from an electronic purse; the security inherent in such schemes ensures the integrity and confidentiality of the transaction. E-purses are ideal for low-value transactions and for children buying on-line.

+ *Virtual cards:* Card numbers, indistinguishable from a standard credit or debit card number but not associated with a physical card, can be used to make payments. The customer effectively passes the payment request to a payment server, which affects the transaction securely with the merchant or the merchant's bank, and then notifies the customer when it has been completed.

+ *Cash on delivery:* In countries such as Japan, Sweden, and Norway, where either customers do not have access to credit cards or a physical signature is required to enforce a contract, e-commerce goods are often delivered to the home or collected by the customer from a corner shop, with payment at the time of delivery or collection.

The companies offering many of these services are not banks. They are information companies, and only some of them use conventional banking services (such as the credit card networks) to clear transactions. The banks see this as a clear threat, but have yet to develop a consistent counter-approach. The banking industry's approach to electronic transactions will be described in Chapter 14.

But smart cards can help in many of these cases: For electronic purses they are essential, and in many of the other cases they form the critical link in the chain, which reliably identifies the customer. Many of these cards will be issued by the nonbank service providers rather than by the customer's bank.

Digital signatures

The identification of the customer is closely linked with the ability to prove that the customer entered into a contract and to enforce that contract afterwards. This is only possible if the law in the country involved allows a contract to be made on-line.

In several U.S. states, in Canada, and in many European countries, a digital signature already has a legal status equal to that of a paper signature. In Britain the relevant E-commerce Bill was digitally signed by the minister.

For digital signatures to be portable, the relevant certificates must be stored on a portable medium, and the only medium which is recognized as suitable for that purpose is a chip card.

In the United States, the need for on-line authentication is now widely accepted, and this is likely to lead to a proliferation of certificates, and to a demand from users for chip cards on which to store these certificates. In Europe, where chip cards are being introduced quite rapidly for payment (see the next chapter), users will demand that authentication be linked to the payment method. So although in Europe chip cards are likely to bring in authentication, and in the United States the process may be the other way around, the medium-term effect is the same: a widespread use of chip cards with both authentication and payment functionality.

References

[1] "Business Information Security Survey," National Computing Centre, Manchester, United Kingdom (annual).

[2] "Computer Crime and Security Survey," Computer Security Institute, San Francisco, CA (annual).

[3] BS7799 (1999): "Information Security Management: Code of Practice for Information Security Management," British Standards Institute, 1999. To be published as ISO 17799.

[4] "A Token Gesture," *Secure Computing*, February 2000, pp. 40–53.

[5] Froomkin, A. M., "The Essential Role of Trusted Third Parties in Electronic Commerce," *Oregon Law Review*, Vol. 75, 1996, pp. 49–115.

14

Financial Applications

In this chapter we look at the ways smart cards are used in financial and banking environments and, in particular, at the way chips are replacing magnetic stripes on bank-issued payment cards.

Bank cards

Functions

We are so accustomed to seeing and using bank cards that it is sometimes difficult to distinguish between them. Traditionally, banks have issued several distinct types of cards.

- *Account cards:* These serve as an identifier only and are typically issued for accounts where no other type of card would be appropriate (e.g., savings and notice accounts). They avoid the need for customers to remember account numbers or for bank staff to search lists.

- *ATM cards:* These are commonly issued to customers who have no credit, including children and new customers. They may only be used in ATMs connected (directly or through an interbank link) to their home-bank network, which means that the transaction is always checked against the current balance on the bank's own computer system.

+ *Debit cards:* These are now the most common cards in Europe. They can be used in ATMs and in retail shops, and they are always linked to a current account. Transactions are posted to the account as soon as they are received; in the case of retail transactions, this may be immediate, the next day, or after processing of a paper voucher. Most debit card schemes now insist on electronic processing, but the transaction is often authorized by another bank or by the card scheme (Visa, Europay, or MasterCard) rather than by the card issuer; this is called *stand-in processing.*

+ *Credit cards*, the oldest form of bank payment cards, allow a customer to pay on credit, up to a preset limit. Credit cards may be used on-line or off-line, and the level of authorization by the issuer is lower than for debit cards. The risk of both credit losses and fraud is therefore greater; thus, the charges to both merchants and cardholders are higher than with debit cards.

+ *Charge cards* (also called *travel and entertainment cards*) may be issued by nonbanks (e.g., retail finance houses: GE Capital is a large charge card issuer), or by banks. They have no explicit spending limits and work on a deferred payment basis rather than rolling credit (the balance must normally be paid off in full every month). Retailer charges are highest for these cards, reflecting not only marketing costs but also the increased risk of default.

Nearly all debit, credit, and charge cards work within the framework of a card scheme or card association (such as Visa, MasterCard, or American Express). The schemes set up relationships between merchants and their banks (merchant acquirers), cardholders and their banks (card issuers), and acquirers and issuers (see Figure 14.1). Merchants agree to accept cards from all issuers within the scheme. Because issuers and acquirers upgrade their systems and cards at different rates, it must be possible for the newest card issued by a large bank in, say, Germany to be accepted and processed by a small retailer, or hotel in Africa or Asia. The need to maintain backwards compatibility is one of the biggest limitations on the speed of technology diffusion in the card schemes.

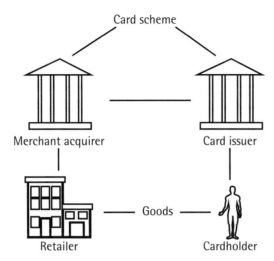

Figure 14.1 Standard model for card payments.

The card associations have reached a very high level of *interoperability*, both internally and between schemes. Cards from all manufacturers and issuers can be accepted by a single terminal at the point of sale. This requires the use of standards, not only for cards but for the electronic messaging systems that transmit and process transactions. The card standards, as we have seen, are based on the ISO 7810 family, while for messaging most countries use a variant of one of the Visa or ISO 8583 standards.

The advent of the smart card allowed banks to move on to the next stage: the *electronic purse*. While credit cards offer "pay later" and debit cards are "pay now," electronic purses are prepaid. The term *stored-value card* is also used in North America, and in Europe when talking about a card which can only be used to pay for goods and services from one source (a vending-machine card, for example, or a multiple-ride ticket for a bus or train service). As soon as it becomes possible to buy goods and services from several different suppliers, this enters the preserve of the banks, and the card is then referred to, in Europe at least, as an electronic purse.

We will also start to see *electronic passbooks*, for accounts that traditionally have been maintained using printed records in a book. Most passbooks carry a magnetic stripe, which duplicates all or some of the printed data in the book, but these suffer from the same security weakness as other magnetic-stripe systems. The use of smart cards will allow the book to be used as a more secure identification of the account holder.

Banks are also seeking to hold more customer data on smart cards held by the customer. This allows them to offer an increased range of services at remote locations where full on-line access would be slow or unfeasible.

Many cards now combine several of the functions described above—debit cards often act as ATM cards and check guarantee cards. In Belgium, Holland, and Germany the electronic purse function is added to the national debit card.

Attacks

The most common source of fraud on cards is theft. When a card is lost or stolen, the thief may be able to use it in an outlet where PINs are not required—particularly if the retailer or shop assistant can be relied upon not to inspect the signature too thoroughly. As soon as the theft is reported, the card may be hotlisted, but this will still only affect transactions in electronic terminals until the next paper hotlist is published. With more than 20 million lost and stolen bank cards worldwide, even electronic lists must be prioritized before transmission.

When the PIN is stolen with the card, considerably greater damage can be done because the card can then be used at ATMs and PIN-enabled merchants.

Thieves can also obtain cards by intercepting them before they reach the valid cardholder. There are various techniques for doing this, but the effect is to obtain an unsigned card, so that the thief can sign it. The valid cardholder is unlikely to know that the card has been intercepted for several days at least, so the time delay before the card is reported lost will probably be longer.

Other attacks on magnetic-stripe cards include the following.

- *Replay:* Merchants swipe the card several times or reproduce the card number or transaction electronically.

- *Counterfeit:* A card that appears to be valid is manufactured (it is often a copy of an existing card) and used for transactions.

- *Card-not-present:* With the increasing acceptance of card numbers over the telephone and the Internet, the card itself need no longer be present when the transaction is carried out. A copy of the card number and expiry date (from an old receipt or from a merchant file) is sufficient to carry out a transaction.

PINs and other electronic checking of cardholder identity protect against several of these attacks, while others can be minimized by the use of on-line transaction authorization. PINs with magnetic-stripe cards are only checked on-line, as they involve the use of a master key. In the United States and many central European countries, on-line authorization is used for almost all transactions, but in some other countries the level of authorization is as low as 20%. Since most fraudulent transactions are authorized, and there is little correlation between authorization rates and fraud levels, a better method of controlling fraudulent transactions must be sought.

During the 1980s and 1990s the major card schemes tested several competing technologies, but by 1995, it was clear that only chip cards offered the right combination of functions, availability, and price, and so a decision was made in principle to migrate from magstripe to chip cards.

Credit/debit cards

Requirements

The prime requirement for a smart credit or debit card is to reduce the opportunities for fraud without increasing the need for on-line authorization. In fact, as banks seek to drive more and more transactions towards cards rather than the less profitable media of checks and cash, the threshold for transactions is moving downwards. As a result, card transactions must be made economical for transactions of a few cents or even less. This creates a firm requirement for off-line transactions.

The first requirement is therefore to be able to authenticate the card off-line. All cards are issued by an issuer that belongs to a scheme. All terminals linked to that scheme must as a minimum have loaded the scheme's public key. This enables the card to be authenticated using a hierarchy: the card itself contains a certificate issued under the secret key of its card issuer, whose public key, signed by the scheme, also features on the card. This form of static data authentication (SDA) can be checked by any terminal loaded with the scheme public key and does not require any cryptographic processing by the card (see Figure 14.2).

It would not, however, detect a card that had been made using an exact copy of the original card data (including the certificate). This requires dynamic data authentication (DDA). When DDA is used, the card itself has unique public and private keys. The card's public key is signed by the card issuer and this certificate is stored in the card. The terminal starts the

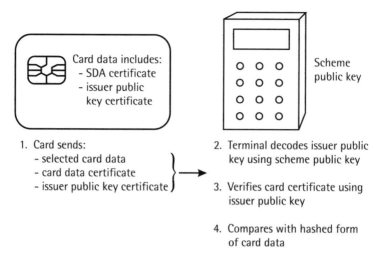

1. Card sends:
   - selected card data
   - card data certificate
   - issuer public key certificate

2. Terminal decodes issuer public
   key using scheme public key

3. Verifies card certificate using
   issuer public key

4. Compares with hashed form
   of card data

Figure 14.2 Static data authentication.

process in the same way, by retrieving the issuer public key, but now it uses the issuer public key to retrieve the card public key. It then sends to the card a message, called the "challenge," containing not only some card data, but also a random number. The card uses its private key to encrypt the challenge and the terminal decrypts it using the card public key. If the two match, then we can be sure that the card really is the genuine card (since by design, no two cards have the same keys). Figures 14.3 and 14.4 show the DDA process.

A further check of the card authentication can be carried out if the transaction occurs on-line. In this case, a symmetric key can be used because the keys do not have to be distributed throughout the

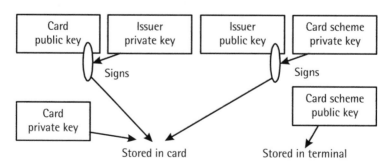

Figure 14.3 Dynamic data authentication—issuing.

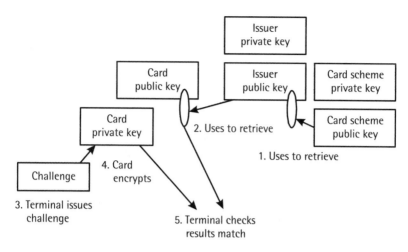

Figure 14.4 Dynamic data authentication—usage.

terminal network. Most bank cards can perform DES encryption within the card.

The next requirement is for verification of the identity of the person presenting the card. This is essential for any unattended environment, and even in retail shops it removes the need for assistants to check signatures or other subjective criteria. Retailers do not like having to perform these checks on behalf of the bank; challenging someone on the basis of their signature or appearance can provoke a confrontation. The range of biometric checks available on smart cards was described in Chapter 6, but the PIN remains the most common CVM by far in banking applications. Where PIN checking is used with smart cards, the PIN may be checked off-line, by the card.

Last, the card issuer seeks to ensure that customers do not exceed their credit limits. With most types of accounts, there are ways of carrying out transactions other than using the card (such as making payments to the account). So the card cannot carry a full record of the current balance. The card is therefore used to force the transaction on-line under certain conditions; these would include high-value transactions, a long sequence of off-line transactions, or a volume limit, which can be customer specific. These limits can in principle be varied dynamically, allowing a very detailed level of risk management.

This model allows card operations to be extended not only to lower value transactions and to unattended terminals, but also to those countries where the telecommunications infrastructure is not sufficiently well developed or where telephone calls are too expensive to permit a high proportion of all transactions to be carried out on-line.

It also offers the possibility of *cardholder controls*. Some customers want to be able to control their spending very tightly and are prepared to put up with some inconvenience or delay to achieve this. For others, convenience and speed are more important than detailed records of all transactions. The smart card allows either of these requirements to be accommodated within a single scheme. Customers may also opt for different levels of control on different cards on the same account; this would allow a parent to give a card to a child, subject to strict limits on expenditure per transaction, expenditure within a day, or on the type of goods that may be bought.

Standards

The need to maintain global interoperability dictated the development of standards for smart payment cards. The card-payment schemes are owned by their members, whose priorities cover a wide range. Four of the five main schemes are headquartered in the United States, where chip cards were long felt to be an expensive way to manage fraud.

In France, however, the smart card became a national technology and source of pride in the 1980s at a time of high and rising fraud. The French banks had a single association (the GIE Cartes Bancaires, known as CB) which controlled card-payment standards. It was tasked with developing a chip-based system, which started trials in 1987. In 1990 the decision was taken to proceed with a rollout and since 1993 all domestic cards have incorporated a chip.

This program has been exceptionally successful; the level of fraud on French-issued cards is less than 0.02% of turnover—less than a quarter of the worldwide average and a tenth of the level at the start of the program. Much of the residual fraud takes place overseas. And the level of on-line authorizations, at under 20%, is also very low by world standards.

Because this scheme was put in place before any other countries had considered adopting smart bank cards, the standards for the French program were created by CB and reflect the structure of the French market and network. A massive migration plan is now under way to convert, first, all French terminals to accept the internationally defined EMV standards, and then to add EMV functionality to French cards so that they can be used overseas in chip mode (whereas today they use only the magstripe outside France).

The Europay–MasterCard–Visa (EMV) standards (see Appendix A) have now been agreed by banks all over the world as the basis for smart credit and debit cards. The EMV standards themselves are concerned with interoperability: they build on the ISO 7816 standards for electrical interfaces and protocols, but give some details in areas left open within ISO 7816. EMV also adds specific functions required for bank cards, such as static data authentication and PIN checking, and it defines the data elements (account number, transaction counter, etc.) that applications may use.

The card schemes have each defined applications using EMV (the Visa ICC Specification and MasterCard Chip Payment Application); they are very similar and can be handled within a single terminal but using different data and keys. Even at this level, however, there is a need for flexibility: one country may insist on using on-line PIN checking for all transactions, while another uses signatures. Or one bank wants to issue multifunction cards with separate settings for Internet transactions and overseas transactions, where its competitor would prefer a simple card with only one set of functions. Table 14.1 shows the elements specified by EMV, VIS/MCPA, and by national standards.

TABLE 14.1 Bank Card Specification Functions

ISO 7816	EMV	VIS/MCPA	NATIONAL STANDARDS	NATIONAL OR BANK OPTIONS
Physical dimensions	Select protocols	Restrict authentication options	Select authentication options	Cardholder verification method
Protocol options	Payment-related functions	Scheme key management	Terminal specification	Additional functions on card
Application selection	Authentication options	Process flow	Host communications	
	Data elements	Scheme rules	User interface	
	Certification process			
Generic intersector card	Generic payment card	Interoperable credit/debit card	—	Bank product

The EMV standards were first published in June 1994; they have been updated several times since, most recently in December 2000. There is a constant struggle between the need to maintain stability and backwards compatibility and the rapid evolution of the technology and marketplace. Arguably, EMV has already succeeded by demonstrating that interoperability is achievable; the longer term evolution of bank products may now take place against that background but without being constrained by the detail.

EMV was defined with credit and debit cards in mind, but an EMV-compatible card may also contain other applications, such as an electronic purse. Although there can be several applications on a card, EMV in its present form does not define any mechanisms for communication or protection between applications—it could be regarded as a single-tasking operating system.

The EMV mechanisms for card authentication include static data authentication, dynamic data authentication (both using public key encryption as described earlier), and on-line authentication (using triple-DES with a derived key). Message authentication is also DES based, and there is provision for PINs, on-card biometrics, or signature checking. The general model of an EMV transaction is described in Figure 14.5.

Banks in Europe and Asia are now moving steadily (albeit at different speeds) towards the implementation of EMV for credit and debit. In the lead is the United Kingdom, which expects to have 85% of its cards and 65% of transactions chip-based by 2003. Most other European banks will at least have converted their systems by the end of 2005; rollout may take an additional two to three years. Asia is up to two years behind this, although fraud "hot spots" will be attacked much more quickly.

North America remains resistant for the time being; as we have argued in Chapter 13, the growth of Internet transactions and of on-line banking is likely to be the driver for chip card issuance in this part of the world.

Procedures

There are few changes in the procedure when a card is used. The card must be inserted into the correct terminal to read the chip (data on the magnetic stripe identifies the presence of a chip). A PIN or biometric may be checked—the PIN entry procedure must be controlled so that only the cardholder sees the number, and terminals themselves must not offer the opportunity to read the PIN electronically as it is input.

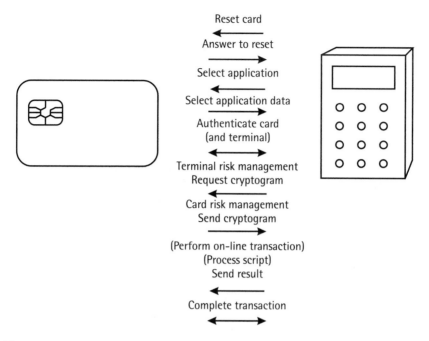

Figure 14.5 EMV transaction model.

Smart card issuers have available a large amount of extra data associated with each transaction. Some of this data may be useful in spotting patterns that might indicate fraud, and the systems used for detecting these patterns on magnetic-stripe systems must be updated. Issuers must also make decisions as to the meaning of different failure conditions—does it mean that the card has been tampered with or could an item of equipment be faulty?

The biggest changes for card issuers, however, are in the areas of key management and risk management. Although all banking standards require the use of host security modules for the generation, storage, and processing of keys, the manual procedures and controls are still important, as we emphasized in Chapter 10. Every financial institution has a data-security officer, a data-protection officer, and a risk-management department, and these people will all be closely involved in the implementation of a chip-based credit or debit card scheme.

Although larger issuers may personalize their own cards, most banks subcontract the personalization process to the card manufacturer. This means that the manufacturer's processes must be bound in to the security and audit processes of the bank. Since a manufacturer will work for several banks, this requires an exceptionally tight control of procedure, data, and in particular keys.

Electronic purses

Requirements

Electronic purses are usually associated with smaller purchases where there is less need for individual controls on each transaction. The procedure for using an electronic purse is usually kept very simple (see Figure 14.6), but this simplicity conceals a process with an exceptionally high need for security.

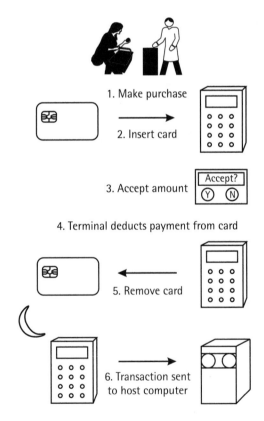

Figure 14.6 Using an electronic purse.

Electronic purses have a special position because of their relationship with cash. If an electronic purse can be counterfeit or the amount on the purse manipulated, this is the same as forging banknotes. Central banks are also concerned about the monetary control aspects of electronic purses and will not allow any commercial business to create tradable value denominated in a national currency [1].

The consensus view is now that:

* Small stored-value schemes are not regarded as e-money and can be operated by anyone within the law;

* Organizations issuing electronic money (whether in the form of cards or software tokens) should be substantial and regulated by a suitable financial services regulator;

* Issuers of e-money must reserve an equivalent amount of funds, so that they are not creating value, but exchanging one form for another.

This view is now encapsulated in the European Directive on Electronic Money [2].

The Bank for International Settlements also considered electronic purses in a report for the G-10 central banks [3] on the security of electronic money. It concluded that purses must use very long keys (128 bits for DES and 2,048 bits for RSA) to be considered secure for many years to come. The Bank also recognized the importance of overall risk management and independent security checks in the design and management of such schemes.

Security is also important for the issuer, who is at risk if extra value or transactions are created. Merchants stand to lose money if transactions (which are usually only held in electronic form) are lost. Both of these are less concerned with the effect on the consumer if the card is lost, but consumers are less likely to hold significant value on the card if they feel that they may lose it all when they lose the card. Some schemes are able to block a lost or stolen card, calculate the value left on the card, and issue a replacement.

Cards are a convenient way to carry money without the need to have the correct coins for a given transaction. They are therefore most popular in unattended applications or on public transportation. These correspond with the situations in which there is no human operator to check the card or cardholder. The relationship of trust is no longer between the retailer and customer but between the card and terminal: the card issues the guarantee of payment, digitally signed by the card issuer.

Those who travel across borders have a far greater problem carrying coins and notes; they end up holding a significant float in different currencies or exchanging cash at a large loss on their return. Few electronic purse schemes can handle multiple currencies today, but the number that can is growing. With the advent of the euro, there is an opportunity to encourage the use of purses rather than notes and coins, and this is being encouraged by the Smart Euro project [4]. For euro currencies, conversion can take place within the purse, using a currency table held in the terminal.

Another area where we can expect electronic purses to be used more widely is for micropayments. Today it is not worth the effort for a company to charge a fraction of a cent for, say, access to a piece of information. The transaction costs exceed its value. The information is therefore either provided free or not at all. If payments can be made electronically, at zero marginal cost, then it becomes worth allowing access to the information at a tiny charge. This principle can be applied to other forms of goods or services; retailers no longer need be limited to charging in whole pennies, for example. Electronic purses provide one possible means to store and exchange these small units of value and to transmit them to a host system in bulk.

Types

The simplest form of electronic purse is the *stored-value card*. We usually take this to mean a card with a single store of value, issued by a company and only usable to buy goods and services from that company. Stored-value cards are used, for example, in vending machines, retail-loyalty schemes, and for cashless payment in clubs and canteens. Their function is almost identical to that of a telephone card, and they can usually make use of the authentication methods used by those cards.

When a card is designed to be used, like cash, in open circulation, then it is considered to be an electronic purse. Most electronic purse schemes involve a value issuer, a card issuer or issuers, merchants, and cardholders. They may also make use of separate acquirers as shown in Figure 14.1.

This means that transactions are asymmetric: The retailer terminal is in control of the card. Transactions are normally made by deducting value from the card, although some schemes will also allow refunds to the card. The cardholder loads the purse at an ATM or special load terminal located in the street or in the home, and typically a PIN is required for the load process.

In most cases, these are the only transactions allowed, and thus the card issuer can hold a "shadow account" which tracks the balance on the card. Any discrepancy can be detected as soon as the transaction is reported.

The simplest electronic purses are, like most telephone cards, disposable: They are loaded before issue and thereafter the value can only be decremented. The more elaborate schemes allow reloading of the purse (some schemes have both types of cards available). Cards can be reloaded at an ATM, through a specially adapted telephone, or using cash at a *card-validation station*.

Disposable cards are almost always anonymous; they are sold by appointed retailers, or company service points, and carry no link with the customer. For cards to be reloadable from an account, they must be personal to that customer. They will either carry details of the account to which they are linked or a reference number which can be linked to the account. In these cases, the card will probably be personalized by printing the customer's name on the card as well.

Security mechanisms

Most of the first-generation electronic purses used symmetric encryption for all authentication and encryption functions. As we described in Chapter 5, this offered a relatively easy and low-cost implementation with standard 8-bit cards. Using this technology, transaction times can be kept down to around 2 seconds. On the other hand, secure access modules (SAMs) are required in each terminal to store the scheme secret keys—the SAM is potentially a weak point in the system and its security is critical. As we have seen, secret-key-based schemes are difficult to administer on a large scale.

So the more modern schemes make use of public-key authentication. Here only the scheme's public key need be stored in the terminal—this is much safer and easier to administer. The drawback is that cards must be more powerful and transactions are slower.

One of the first public-key purse schemes was Mondex. The Mondex scheme is unique in that it also allows *card-to-card transactions* which do not pass through a merchant terminal, but instead through a special wallet which also contains a Mondex purse (Figure 14.7). This allows cardholders to make interpersonal payments to members of their own family, friends,

Figure 14.7 Mondex wallet. (Photograph courtesy of Mondex International Ltd.)

or casual service providers who are not card-scheme merchants (e.g., domestic staff or window cleaners).

This feature of Mondex has been criticized by some bankers because it does not allow checking and audit of every transaction. Value can in principle circulate for long periods without being cleared through a bank. Although the incompleteness of the record does demand a higher level of security and, as we shall see, some additional checks, the Mondex model is proving popular in semi-closed environments such as universities, and is still being tried by several banks around the world.

CEPS

Although most purse applications work in a restricted national or application environment, there is a perception that e-purses will only take off if cards can be used everywhere, even across borders. In Europe, it is against the principle of the single currency to have a product that cannot be used throughout the euro area.

This problem was addressed by the European Committee for Banking Standards in 1996–1997. The ECBS recognized that there were too many different purse schemes in existence to impose a new standard on all, so instead opted for a set of common functions which could sit alongside an existing purse scheme, or could be used to form a purse scheme on their own.

This work resulted in the development of the Common Electronic Purse Specifications (CEPS). CEPS uses public-key authentication and symmetric-key encryption, in the same way as EMV credit/debit cards. The low-level electrical, protocol, and application selection functions are the same as for EMV, but the application functions (load, purchase, refund, view balance, etc.) are defined by CEPS rather than EMV.

Most of the electronic purse schemes currently in operation have expressed their support for CEPS, although there are still relatively few full implementations available.

Status

Most European countries, and many others around the world, have at least trialled an electronic purse scheme. Banks in some countries, notably Germany, have gone so far as to issue e-purse cards to all their account holders. But usage of the cards is consistently low, with average transaction rates rarely higher than two or three transactions a year.

A number of reasons have been put forward for this, including pricing, attractiveness for merchants, and consumers' preferences for anonymous cash [5, 6]. The technology itself has rarely been called into question, although some schemes suffer from slow transactions.

Most current schemes belong to one of three or four "families" of licensed designs:

- *VisaCash* is an umbrella name for a group of schemes that were not all compatible. VisaCash schemes being launched after 2000 will conform to the Visa CEPS standards.

- *Proton* cards were originally developed in Belgium, but are now used in over 20 countries. The original technology is a single-currency, symmetric-encryption scheme, but Proton World has demonstrated a CEPS application running alongside the Proton application. The Belgian scheme remains the most intensively used of all purse schemes, with up to 0.5 transactions per card per month.

- *Mondex* was described above. Its unique features make it incompatible with the CEPS protocols, but it still attracts considerable interest because of its potential for minimizing transaction costs. Mondex will normally be implemented on a Multos card (see Chapter 9).

 • *Mifare* cards are contactless, and are primarily used in public transportation. But they also have simple e-purse functionality.

Table 14.2 shows some of the schemes currently in operation and their main features.

On-line transactions

On-line banking

The most common type of on-line financial transaction is on-line banking: viewing an account, setting up or initiating transfers between accounts, and in some cases sending payment instructions are the most common functions.

Early on-line banking systems used private networks, accessed through a special number. However, since the public switched network was involved this offered little extra security compared with the Internet, which is now the normal form of access.

Clearly authentication of the customer is critical in this environment, but most current schemes for personal customers only use passwords—often long and complex password sequences which mean that customers are more likely to write them down. Corporate banking systems are much more likely to use a smart card or other token.

In the Nordic countries, where Internet banking is highly developed, banks have opted for a *one-time pad* form of authentication. Customers are issued with a list of 20 or so numbers, and each time they access the system they use the next one on the list, striking it off as they do so.

The German banks have opted for a chip card–based scheme for authenticating customers and signing transactions. The Home Banking Computer Interface (HBCI) meets the requirements of the German Digital Signature law, and there are plans to use it for all communications between individuals and public authorities. To date around 400,000 HBCI cards and readers have been issued, out of 6 million home-banking users. The user is authenticated to the card using the one-time pad described above.

Transaction authorization

Any type of transaction, such as sending a purchase order, placing an advertisement, or making a payment, can be authorized using a smart card. A public-key system is used, with the secret key held on the card and the public key made available to the other party. A digest of the message is

TABLE 14.2   Selected Electronic Purse Schemes Operational in 2000

Country	Scheme Name	Status (mid-2000)	Features
Australia	Ecard	Operational (800,000 cards)	Combined health and purse card; Proton model
Austria	Quick	Operational (3.5 million cards)	—
Belgium	Proton	Operational (7.5 million cards)	Reloadable cards; terminals require SAM
Canada	Mondex VisaCash	Pilot Sherbrooke Pilot Barrie, Ontario	—
Chile	Supercard	Pilot (500,000 cards issued)	Proton model
Denmark	Danmont	Fully operational since 1994: 400,000 cards	Disposable and reloadable cards
Finland	Avant	Being rolled out: 400,000 cards at end of 1999	Disposable and reloadable cards
Germany	Geldkarte	Function added to 50 million national debit cards	Proprietary technology
Italy	MINIpay	Large-scale pilot: 1 million cards	—
Japan	Mondex	In-house scheme for Hitachi (100,000 cards)	Staff ID and e-purse on a Multos card; building access using Mifare application planned
The Netherlands	Chip Knip Chipper	Operational (14 million cards) Operational (13 million cards)	Proton model
Portugal	MEP	Operational (400,000 cards)	Proprietary technology: reloadable cards; uses SAM
Russia	Zolotaya Korona	Several regional schemes operating	Reloadable; multiple currencies
Singapore	Cashcard	Operational (3 million cards, 84 million transactions in 1999)	Cashcard is the only way to pass the tollbooths to access the city center
Spain	VisaCash	Operational (4.2 million cards)	—
Sweden	CASH	Being rolled out (2.5 million cards)	Proton model
Switzerland	CASH	Pilot terminated in April 2000	—
United Kingdom	Mondex VisaCash	Operational in several university campuses Pilot terminated in June 2000	Off-line/card-to-card operations
United States	Mondex	Internet pilot in San Francisco	—

created by the software; this is signed by the card using the secret key and checked by the recipient using the public key.

If the process is repeated in reverse (when the transaction is acknowledged), then the originator is able to authenticate the recipient. This mechanism provides a complete check of the integrity of the message and neither party can subsequently repudiate the transaction.

In the United States, American Express has led the way in offering a card with this form of authorization for use in on-line transactions. The "Blue Card" was the first mass-market financial smart card in the United States to combine the functions of a credit card, X.509 digital signature, and a server-based wallet for on-line payments. With a server-based wallet, the user authenticates him- or herself to the server, which then carries out the transaction in a secure mode, directly with the merchant or merchant gateway. Server-based wallets may be operated directly by the card issuer (see Figure 14.8).

Secure electronic transactions

As we saw in Chapter 13, e-commerce transactions are a headache for banks, because few existing mechanisms have the penetration and security required. The rate of disputed transactions is at least 10 times as high as for face-to-face transactions.

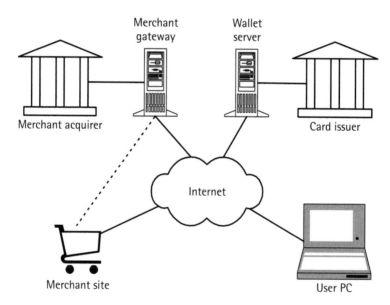

Figure 14.8 Server-based wallet operation.

There is a need not only to authenticate the customer (ensure that they are not simply using someone else's card), but also to store a complete record of the transaction (what goods were ordered, any delivery charge, how the customer expressed their agreement), which can be accessed in the event of a dispute. Recurring payments (e.g., monthly subscriptions) cause particular problems, which can take months to resolve—it is often very difficult to prove just what the customer agreed to.

The use of SSL and other channel encryption protocols does not resolve this problem, since customers do not in general have certificates and are unable to check merchant certificates. And many of the problems relate to transactions which all parties agree did take place—it is the details which are being queried.

In order to resolve this problem, the major card schemes developed the Secure Electronic Transactions (SET) protocol [7]. Cardholders must load into their PC a SET wallet and a certificate issued by their card issuer. Merchants have special SET software and a certificate issued by the merchant acquirer. The transaction details are encrypted and signed by both parties—the merchant never sees the card number and the bank never sees the transaction information. But the overall signature on the message allows all parties to check that the transaction did take place, that it was authorized by the cardholder, and that the details were as claimed by the merchant (see Figure 14.9).

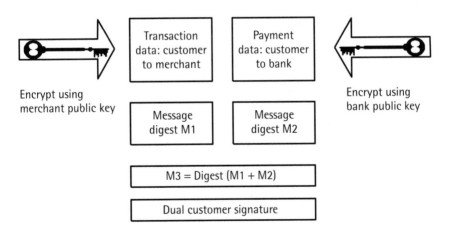

Figure 14.9 SET message.

SET is a very secure solution, but it ran into a number of problems:

♦ The cardholder wallet was large and slow to download;

♦ Customers had little incentive to use SET because few merchant sites accepted it, and merchants would not gain sales because few cardholders had certificates;

♦ Transactions were very slow: 40–45 seconds was the norm.

SET is improved by the use of server-based wallets and certificates, as shown in Figure 14.8. This overcomes the problems of distribution and speed, but there is still a difficulty in that existing sites are seen to operate successfully with SSL—only a very small proportion of merchants causes the problems referred to above. Many large merchants have bricks-and-mortar operations as well, and are unwilling to use different protocols for their Internet transactions. In the United States in particular, banks felt that enforcing the use of SET would drive merchants to alternative payment providers.

The solution now favored is to separate the card issuer, acquirer, and interoperability domains (see Figure 14.10). Card issuers can use server-based certificates and authentication methods, which may be accessed by other channels such as mobile phones, or which may be the same as current on-line banking authentication. Acquirers can use different protocols according to the merchant type. And the issuer and acquirer servers can negotiate a mutually acceptable protocol, which might be SET in Europe or SSL in the United States. In the longer term, IP v6 will also meet the requirements of the interoperability domain, as would the PKIs being set up for other clearing systems.

Where banks have already issued chip cards to customers, this will probably form a major part of the authentication in the card-issuer domain. For example, the French banks are cooperating within a project known as Cyber-Comm, which aims to sell cardholders a special card-reader with PINpad and display, for checking PINs off-line before authorizing transactions.

Benefits payment

Fraudulent applications and collection of Social Security and unemployment benefits are widespread in every country. A system that permits

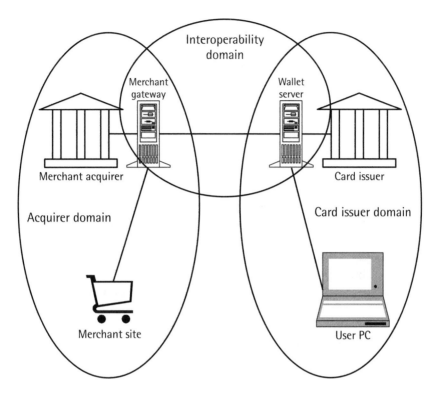

Figure 14.10 Three-domain model.

positive identification of claimants offers large savings. Even in countries such as the United Kingdom or the United States, where there is no unique register of citizens, details of claimants (including a biometric) can legitimately be stored and duplicates can be detected.

There have been several pilots of smart card–based benefits collection, some based on signature verification and others on fingerprints. Fingerprints are preferable in areas where literacy rates are low, and this is a case where security takes priority over customer preferences.

Where a smart card is used for identification, it also offers a payment mechanism: the card itself can be an electronic purse, or the amount loaded may be transferred to a bank account by inserting it into an ATM or telephone terminal. The paying agency saves cash-handling costs, and there is an immediate and complete audit trail.

This model also allows control over the way in which a specific benefit is used. For example, some schools in Britain use smart cards to provide a free school meal allowance to pupils who are entitled to this benefit. A food benefit could be restricted to food stores or a mobility allowance could be restricted to public transportation or gas costs.

The tools required to implement such a scheme are no different from any other electronic-purse system, while a stronger CVM is available through the biometric.

Loyalty

Many retailers have recognized that retaining and building the loyalty of customers is cheaper than attracting new customers. One of the main tools used to this end is the loyalty card. Customers collect points on their card every time they make a purchase, and this can subsequently be redeemed for goods in the store or from a catalog.

Loyalty schemes allow the retailer to collect additional data about customers' purchasing habits, but they are often expensive both in terms of margin (the value of the points offered) and administration costs. Some schemes operate on-line (all points are recorded on a central system), while others store the points on a card. Off-line points storage on a magnetic-stripe card is subject to fraud, and the storage available on the stripe limits the amount of data that can be collected.

Smart card loyalty schemes help to overcome these problems: the card can include authentication functions or encrypted data to prevent counterfeit and alteration. Redemption can be controlled by using a PIN. Although points will usually be issued by an on-line electronic point-of-sale (EPOS) system, the level of points issued and the message displayed to the customer can be varied by using the customer data stored on the card.

Smart cards can also be used as staff cards or discount cards in a single consistent scheme.

New delivery channels

The PC and Internet have now become standard delivery channels for banking services. They are attractive to banks because the cost of delivery is much lower than with branch-based, or even telephone banking. But although the proportion of the population with Internet access is growing everywhere, penetration is still much lower than that of the television or mobile phone.

The use of chip cards opens up the possibility of delivering banking services and allowing payment through cable and satellite television, mobile phones, and personal digital assistants (PDAs), kiosks, and screenphones. Again, the use of server-based applications and authentication through the chip card give the flexibility needed for these developments.

References

[1] *Report to the Council of the European Monetary Institute on Prepaid Cards,* Working Group on EU Payment Systems, May 1994.

[2] *Electronic Money Directive,* European Commission, July 2000.

[3] *Security of Electronic Money,* Report by the Committee on Payment and Settlement Systems and the Group of Computer Experts of the Central Banks of the Group of Ten Countries, BIS, Basle, 1996.

[4] *The Euro in the Electronic Purse,* Report by the Smart Euro Partners for Eurosmart, April 2000.

[5] Hendry, M., "The Right Strategy," *European Card Review,* July/August 1999.

[6] Birch, D., and M. Hendry, *Retail and Consumer Payments in Europe and North America,* London: Informa Publishing Group, 2000.

[7] Loeb, L., *Secure Electronic Transactions,* Norwood, MA: Artech House, 1998.

15

Health

Healthcare and medical cards are probably the largest application of smart cards in which an older technology is not being replaced. Healthcare is a highly political subject, and there are wide variations between countries in the extent to which health cards are feasible. Countries also prioritize the applications of the cards in different ways. Nevertheless, the use of smart cards for such a wide range of applications in this sector suggests that there is a common need for the security and storage that only these cards can give.

Insurance

The dominant motive for introducing health cards is control of costs. Although the structure of health insurance varies widely from country to country, each has a common requirement to verify that the person claiming medical services is insured, and to correlate claims for payment with both individual patient records and the doctors' accounts. The majority of cards in the health sector are therefore insurance cards without any medical application.

Health insurance cards do not usually contain a sophisticated card-holder identification method; if they do not contain any medical data, they may well have no cardholder verification at all. The confirmation of identity is expected to be by a separate method, typically personal knowledge of the patient by the doctor or presentation of a national identity card.

The *Versichertenkarte* in Germany is issued to every person covered by health insurance. All of the main insurers participate, and virtually the whole population of 80 million now carries a card. The main security device on these cards is the general use of write-protected memory; the card can be read in any authorized hospital, clinic, or general practice, but it can only be written to by the insurer. The cards also contain an authentication area that allows each insurer to identify its own cards [1].

In France, the *Sesame Vitale* scheme has issued over 40 million cards to patients and 225,000 cards to doctors, pharmacists, and other health professionals. Ultimately the system is intended to cover the whole French population covered under the compulsory national insurance scheme. The Sesame cards, which are issued by the Caisse Nationale d'Assurance Maladie (National Health Insurance Scheme) serve both to identify the cardholder to the doctor and to provide proof of insurance. Medical records are held separately, although the card is also used to convey prescriptions from physicians to pharmacists. Each card covers a family or *insurance unit*.

Every terminal dials in regularly via the national X.25 network to the social security office to record the services performed or prescriptions dispensed. These records can then be correlated with the claims received from patients. The main aim of the system is to reduce paperwork, but it also contributes significantly to security, reducing the scope for fraudulent claims, and ensuring a double-ended check on every transaction.

In the United States, several insurers and health maintenance organizations (HMOs) are now starting to use smart cards for patient and claim verification, and several of these say that they intend to move on to also storing medical records on the card [2].

Medical records

Alternative approaches

Medical records involve large quantities of data. A single patient's records will amount to several hundred kilobytes over a lifetime, and for those with chronic conditions this can extend to megabytes. Storage of this magnitude is probably not economic with current smart card technology, so several alternative approaches have been followed:

- Storing the data on an on-line system, using the card to control access to the records;

- Keeping only essential summary information on the card, with references to other records held on-line;

- Using the card for specific conditions or applications (e.g., hospital treatment or long-term tracking of a condition);

- Restricting the card's application to that of a short-term data carrier, holding only current data and communications between health professionals (e.g., referrals or prescriptions);

- Using a smart card in combination with another card or storage medium to carry the data.

The first of these options (full on-line storage) is unlikely to be practical in anything other than a very small community. In most countries, patients move freely from one healthcare provider to another, using many different services. One of the main attractions of the card, therefore, is that it can be held by the patient and used wherever he or she goes for treatment. In Europe, patients can and do move freely from one country to another, and so the long-term requirement is for a system that can be applied throughout the European Union.

The second option (storing only the minimum summary data on the card) is more widely used; this would alert a doctor to special conditions and other treatment that could affect the diagnosis or recommendation. Each of the other approaches has also been used, and each has its own advantages and drawbacks.

Issues

The main issue for a medical data card is the need to protect the *confidentiality* of the data stored on the card and to *restrict access* to qualified health professionals. A common feature of nearly all such schemes, therefore, is the existence of a professional card as well as the patient card. Both cards must usually be inserted into the terminal before the data on the patient card can be accessed. Many schemes require the ability to collect or enter data through a portable terminal (for example, on a home visit or during a hospital round). Special terminals are therefore a common feature of these schemes.

Another problem common to all medical applications of smart cards is the need for *patient consent*. Many people want to understand what is being held on file about them and to control who sees that information. This is not as simple as allowing patients to view the data on their cards; the data will often be in codes and use words not understood by the patient, which may cause unnecessary alarm. A general practitioner will often want to explain to a patient the implications of the data recorded, and this can be time consuming.

With a health card, patients may control who sees the information simply by deciding whether or not to hand over the card. If the card contains administrative as well as health data, then it will be necessary to protect the confidential health data using a PIN or password.

Patient-doctor confidentiality must be carefully maintained. In many cases, the doctor may pass on information to another professional on a need-to-know basis, and generally the patient's consent is not sought or required for this (it is regarded as implicit when the patient agrees to consult the other physician), although most doctors will seek the patient's agreement in the case of particularly sensitive information. Using a smart card or any other electronic system formalizes these relationships and the need for consent, and sometimes removes an element of discretion that can benefit both doctors and patients.

Many of these issues are seen as less important when treating emergencies or chronic illness, and so disease-specific (e.g., cancer care) or emergency (blood group and allergy) cards are easier to implement.

No system can operate without agreement on the *standard codes* to be used for diseases, diagnoses, and drugs. Considerable efforts are going into standardizing the use of the International Classification of Diseases (ICD-9) and similar codes. Doctors, clinics, and hospitals use many different computer systems, often designed for the specific requirements of a national healthcare framework. It will be some years before these are brought into line with the current standards. Large gaps also exist in the chain where some providers have no computer system at all.

Not all medical records can satisfactorily be expressed in terms of codes and dates. It is often necessary to accommodate *additional data*, including text and images. Some of the text data may be quite subjective and yet important to the diagnosis; these records are often sensitive in the context of the relationship between physician and patient.

Another problem with patient cards arises if the patient *loses* the card or it is damaged. It will probably be possible to reconstitute the core personal data on the card (in most cases it will be issued by a central authority or "home" medical practice that has access to the central records), but the other data on the card may have been stored by several practitioners using different systems at different times. The record of use of the card is lost, including, for example, a patient's record of drugs dispensed. Some systems therefore copy all data to a master file whenever the card is used in its home location.

Overall, the issue of health cards has political and legal dimensions; as well as having to meet user and acceptability criteria, issuers must be sensitive to these additional dimensions. These are discussed in legal terms in [3].

Operational and pilot schemes

Many technologies have been used for health cards. In Japan and China there are large-scale systems using optical cards (see Chapter 4). The large capacity and write-once-read-many-times characteristic of these cards make them particularly suitable for use in medical-records applications.

The European Union has commissioned several large-scale medical-records card schemes using smart cards, including:

- *Cardlink:* This is an ongoing project involving the issue of around 250,000 "Eurocards" as health passports to patients in nine European countries, for use in medical emergencies. The information on the card can be accessed and updated by general practitioners, hospital specialists, pharmacists, and authorized administrative staff. The aim of the current phase is to test the interoperability and patient acceptability.

- *Diabcard:* This project concerns a patient card for chronic illnesses. This project is also moving into a second phase in which interoperability and user reactions are being tested; in the first phase patient acceptance was high but some physicians were uncomfortable with the limited data on the card.

In a large-scale French trial, 50,000 patients in the Loire region have been issued with *Santal* cards. This is a memory card with several datasets; each dataset can only be accessed by a healthcare professional using the appropriate professional card for that dataset, and only approved professionals may alter or append data in specific fields. The data are divided into:

- *Emergency data:* Corresponding to the Cardlink card;

- *Hospital data:* Admissions and outpatient consultations;

- *Medical data:* General practitioner or physician consultations, diagnoses, and treatments;

- *Laboratory data:* Blood work;

- *Pharmacy data:* Prescriptions issued and dispensed;

- *Nursing care.*

In the Republic of San Marino (25,000 inhabitants), a smart card issued by the local bank as an ATM and debit card contains a special *health zone,* which can be read by the local hospital, clinics, pharmacists, and medical practices.

A large-scale pilot of a hybrid *opto-smart* card is being carried out in Leverkusen, Germany. This is a single card carrying both an optical memory area and an ISO-standard multifunction smart card chip. Administrative and short-term data records are held on the chip, as are the keys and secure authentication protocols. Most of the medical data are held on the optical memory, but in an encrypted form [4].

The number of optical readers is minimized in two ways:

- Pharmacists and emergency departments need only read the smart card, as all the data relevant to them are stored there.

- The optical card is read once when the patient arrives at the clinic or hospital, and the data are made available (still encrypted) on the internal network. The patient grants the doctor permission to read the data by entering a password; only the combination of the patient's password and the doctor's professional card allows access to the encrypted data.

The Leverkusen pilot system also features a site card, which is used to initialize the network server and to authorize a professional to use the network, as well as a certification authority (the *trust center*), which authorizes the issue of all of the cards in the system.

Prescriptions

Several of the schemes already described include a facility for transmitting prescriptions from doctors to pharmacists. There is an increasing need to control access to drugs, both for cost and health reasons, and traditional handwritten paper prescriptions are no longer considered adequate for this task. Printed prescriptions are a slight improvement; however, for full security, an electronic prescription that can be accompanied by a message authentication check is the minimum satisfactory solution. A digital signature could also be added, but this would require a full-scale certification authority arrangement, which has not yet been implemented by any large country.

In the longer term, the certification of health professionals as a part of their qualification process would prevent many cases of fraudulent practitioners and bogus qualifications, and it would permit the implementation of a nationwide or international digital signature scheme.

Almost any smart card, even a small memory card, would permit the operation of a simple prescription-handling system; in addition to preventing prescription fraud, it would form the basis for a control system with a double-ended check on the issue and dispensation of prescriptions.

Patient monitoring

As a further leg in any strategy for controlling costs, the health industry is also seeking ways for patients to monitor themselves to avoid many routine visits. A wide range of machines is available for measuring blood pressure, pulse, sugar levels, and many other indicators in the home. These can often be connected to a home computer or other device; some already have smart card slots.

Patients on regular drug programs, not all of whom can be relied upon to take their drugs regularly, can be prompted by devices that will dispense the drug, prompt the user to take it, ask for confirmation that it has been taken, and record the whole operation on a card.

Those for whom exercise is a recommended treatment can have their programs loaded onto a smart card; the card is inserted into an aerobic exercise machine (a bicycle, treadmill, or step machine) that also features a pulse monitor. The machine then adjusts the work rate to match the program and records the work carried out.

In each case, the data may be transmitted to the doctor monitoring the program by modem or the patient may take it when visiting the doctor. At this stage, such schemes are nearly always voluntary, and there are few security implications. In the longer term, data on these cards should be subjected to similar criteria to those applied to other medical records; although the information is usually less sensitive, the same data protection principles apply.

In general, the data should only be viewed with the consent of the patient by a health professional qualified in the appropriate discipline. The consent may usually be implicit (when the patient hands the card to the doctor or transmits the data to his or her office) or, in the case of more sensitive data, explicit (by entry of a password at the time the data is viewed).

Where such data is transmitted by modem, the link should be established by an authentication routine between the card and the central system (the doctor's surgery or hospital). For maximum security, all data to be transmitted should be encrypted by the card or modem, but this may be considered an unnecessary overhead for this type of data.

References

[1]   Schaefer, O., "Introduction of Chip Technology to Health Administration and Medicine in Germany," *World Card Technology*, Vol. 1, No. 3, 1995.

[2]   Ognibene, P. J., "Card Smarts: Smart Cards Are Changing the Face of Health Insurance," *Technology Decisions*, July 1999.

[3]   Ducrot, H., "Information Médicale: Aspects Déontologiques, Juridiques et de Santé Publique," *Informatique et Santé*, Vol. 8, Springer-Verlag, Paris, 1996.

[4]   Hartmann, G., et al., "Security Scheme for Hybrid Opto-Smart-Health-Card," presented at *Towards an Electronic Patient Record*, San Diego, CA, 1996.

16

Transport

In the modern world, there is so much emphasis on mobility that every system must address the needs of the mobile user. This chapter is concerned with those applications that make mobility itself more efficient. It is the one area where contactless cards have been consistently preferred over other types.

Local public transportation

There has been a massive explosion in chip card–based public transport schemes in the last few years, largely driven by the availability of reasonably priced contactless cards.

Organization

In many large cities, all buses, trams, and underground trains are controlled by a single authority, which may be publicly owned. In most smaller towns and rural areas, and increasingly often in cities as well, the system is run by a network of smaller operators, all privately owned and responsible for their own profit and loss. Even within the large metropolitan authorities, many lines or services are run as profit centers and must account for their income and volume of passengers as well as their costs.

Whichever system is used, an efficient public transportation service requires seamless journeys, with passengers traveling on a single ticket across several operators' services. It may also be necessary to have some agreement with a national or local railroad operator if it provides complementary services. All of these systems bring with them the need to apportion revenue between operators fairly.

The system is made more complex in that public transportation is seen as a social service: Some services may be subsidized, either directly or through license conditions. Elderly passengers and children receive concessionary fares. The result is a very complex system, difficult to control. Enforcing fare collection is often uneconomic, which leads to fare evasion. In many areas, both usage and service frequency are dropping.

Magnetic-stripe cards have been used as tickets on all forms of public transportation for many years. They give little management information, and higher value tickets such as annual season passes are a target for fraudsters—most magstripe- and paper-based systems are reckoned to suffer fraud losses of between 5% and 15% of revenues [1]. The machines for reading magnetic-stripe tickets at high speed are expensive to purchase and to maintain. However, the strongest motive for introducing smart cards as public transportation tickets is the extra opportunity they give to collect usage data and to use this to apportion revenue (see Figure 16.1).

There are now many systems operational in which smart cards are used as season passes and for regular users. Some of these form part of *town card* schemes or other multiapplication cards run by a single operator, but the more advanced and complex schemes are usually restricted to the public transportation function.

For the scheme operator, they offer many benefits: Reducing or eliminating the need for cash collection improves security, both for drivers and for the scheme. There is no delay while drivers check or sell tickets. And it is relatively easy to set up fare schemes that reflect the operator's priorities (including, for example, higher fares for night buses or during peak periods).

Types of card

Most public transportation operators prefer a contactless card, primarily for speed of boarding. When people are in a hurry, they become impatient at the time required to insert a card and carry out the transaction; they

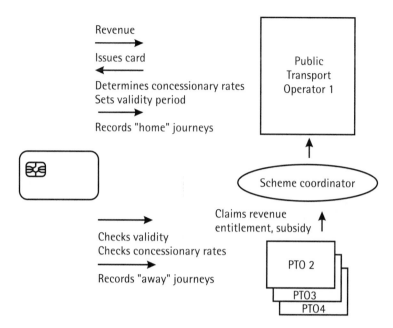

Figure 16.1 Public transport payment and reconciliation.

often remove the card too quickly or damage it physically by mishandling it. With contactless cards, the number of passengers who can board in a given period is much higher. As was mentioned in Chapter 7, contactless readers may also offer higher reliability and ease of cleaning.

Even at the high data transfer rates available with modern contactless cards, the transaction time is still limited by the need to power up the card and establish initial communications. There might also be a need to resolve conflicts between two cards within range of the antenna at the same time. It is found in practice that a 200-ms transaction time is too slow; many passengers have already moved their cards out of range during that time. A realistic maximum seems to be 120–150 ms.

Although reading distances up to 50 cm are possible, most systems use 10 cm or less, and many operators prefer to have passengers bring their card into contact with the reader. This more positive action, confirmed by an audible signal from the reader, results in fewer failed transactions.

The Asian cities of Seoul (Korea) and Hong Kong pioneered this technology. Hong Kong's Octopus system now has 7.5 million cards in issue and handles as many journeys a day. Singapore is installing a similar system. Other European and American cities, including Berlin, Rome, Paris, and San Francisco are now working on chip card systems. Gothenburg in Sweden is one of the first to have an integrated system combining magstripe for low-value single tickets with chip cards for season tickets.

As we saw in Chapter 7, the most common cards still fall into one of two more or less proprietary families (represented by ISO 14443 types A and B), but both systems have now been proven in large-scale schemes and meet the needs of most public-transport operators, at least for self-contained schemes.

Type A represents the MIFARE protocols. The original MIFARE card used an application-specific integrated circuit (ASIC) manufactured by Philips, but cards using this protocol are now available from several other manufacturers and using microprocessors. MIFARE allows mutual authentication between card and reader, encryption of the data being transferred over the RF interface, and an embedded unique serial number per card. The reading system must also incorporate the same proprietary interface.

Type B is used by the ERG/Motorola consortium, and is a development of the Proton electronic purse architecture described in Chapter 14. Both type A and type B cards are now available as *combi cards*, with both contact and contactless interfaces.

Security and interoperability

Security on the public transport schemes described above is ensured by the use of a cryptographic certificate stored on the card and secure access modules in all readers and system devices. This limits the system to a single operator or group of operators that trust each other.

In practice, many public transportation users are mobile: they may visit several different towns or cities. Public transportation operators will usually agree to honor one another's cards provided that they can ensure that revenues can be apportioned correctly. Such agreements could even extend across borders.

There are two ways to design a scheme involving several operators and a common card: either the card can hold multiple counters, one for each operator, or each transaction is recorded by the accepting terminal and all reconciliation is carried out on a central system.

The first method involves lower overheads because the card does most of the accounting and less data needs to be transferred to the central system, while the second keeps a more complete and up-to-date record. In the United Kingdom, the Integrated Transport Smartcard Organization (ITSO) has opted for the second route. It has developed a specification that allows transport operators and similar organizations to read data from tickets issued by other ITSO members, together with a central clearing system for exchange of data.

The advantages of control and savings on fare collection are greatly increased if cash and printed tickets can be eliminated altogether, making the whole process automatic. Many public transportation users, however, do not travel regularly on the same routes and indeed may only be visitors to the locality. The system must be able to handle single ticket buyers as well as season tickets and regular users.

It would be ideal if the transport terminals could accept "open" electronic purses of the type described in Chapter 14. At present, however, there seems to be little likelihood of such a close cooperation between banks and transport companies, and it is more likely that the transport cards will be used for other purposes.

Most of the schemes to date have been either in a single, centrally controlled municipal authority or short-running pilot schemes. Those now being implemented, notably London Transport's Prestige scheme, will have to contend with the longer term issue of the stability of the network of operators.

All of the schemes described here work in a "web of trust" encompassing all of the transportation authorities in an area. This is fine as long as the companies are stable and the area is well defined. The main problems arise when companies join or leave the area, when new contracts are established with an additional company in the area, or an operator goes out of business (possibly with debts to the system).

To overcome this, many schemes work on defined time periods: typically one, two, or three months. They attempt to collect all data within, say, five days of the end of the period, and to complete reconciliation within an additional three days. In this way exposure is limited and a new company can join at the beginning of the next period.

Taxis

Modern taxi fleets have radio terminals using private mobile radio frequencies with encrypted transmission of data between the taxi and central system. This is used for dispatching bookings, bidding for bookings, recording positions (either manually or using the satellite-based GPS), and for making payments.

Regular users are frequently issued with account cards; these not only allow the journey to be charged to the account, but may also specify a cost code and identify the specific user on a corporate account. Bank-issued credit cards can also be accepted; the transmission of the card number to the central system is secured using the digital radio system, and the transaction can be authorized on-line from the central system. Smart card transactions, whether credit, debit, or purse, can be handled in the same way.

Trains

Railroad systems vary considerably around the world in their level of integration and central control. Since the wave of privatization in the 1990s, some countries' train systems have the same problems as those described for local public transportation. In this case, however, the much wider range of fares, coupled with the need for connections and fares for specific destinations, make the whole process still more complex. Tickets are normally still inspected by a ticket inspector, so it is likely that printed tickets will remain with us for many years yet.

The first areas where smart cards may be seen are for season tickets for commuters; here, the requirement is to allow ticket checking without any delay, and a contactless smart card would provide the answer. Such trains have much in common with metro systems in cities, and the same technology can be used.

At the other end of the scale, smart cards may appear for longer journeys in first class and special trains. Here the requirement is not so much for ticketing purposes as for the inseat entertainment systems currently being designed. These will follow the pattern of inflight entertainment, which is already well established.

Air travel

Requirements

Air travel offers several opportunities and challenges for smart card vendors. This is a high-profile sector; frequent fliers are a high-earning group, and airlines must expend considerable effort buying and retaining their loyalty. Nearly all have bank cards, and an increasing number will have smart payment cards. International travelers are frequently faced with foreign exchange problems, having to make payments in a currency they do not normally hold. And the whole air-travel experience is seen as a stress that can be minimized by the use of technology.

Against this, the nature of international business means that vendors have to try to meet national requirements in many areas where no international standards exist. The high values involved, the intrinsic vulnerability of the physical controls, and the ease of escape makes airports a magnet for thieves and fraudsters. Systems must be designed in the certain knowledge that they will be attacked.

Electronic ticketing

The first major application of smart cards in air travel seems to be for fast ticketing or ticketless travel.

Airlines have been experimenting with electronic ticketing for some time, but it was only at the end of 1996 that the International Air Transport Association (IATA) agreed to a set of standards for ticketless travel using either magnetic-stripe or smart cards [2]. This will not only allow ticketless journeys involving more than one carrier, but will also provide the incentive for major airports to equip themselves with electronic check-in desks.

The electronic ticket allows remote downloading of a ticket (which could be over the Internet), unattended check-in (a boarding pass must still be issued to meet Warsaw Convention rules), and passenger tracking through the airport. The same smart card may be used to access inflight entertainment services, book hotels and taxis on arrival, and to record expenses when on a trip.

Several airlines are now piloting smart card electronic-ticketing schemes, often combined with their frequent-flyer programs. These work well, but the level of investment in infrastructure required has so far been a barrier to their wider use.

Inflight entertainment

Many airlines now offer a wide range of entertainment and other services at the seat on long-haul flights. These may include videos, flight information, telephony, purchases, electronic games, and even gambling. Many of these services require payment, and the preferred form of payment from the airlines' point of view is a bank-issued credit or debit card. In some cases, the airline has set up a link between its frequent-flier card and a specific bank card; in these cases the airline would prefer that the customer used its own card—this might even be the only card allowed for some functions.

Current handsets will only accept magnetic-stripe cards, but the latest designs are already taking into account the need for smart card reading. PIN entry, if needed, would require the handsets to be designed and manufactured to a higher standard to maintain the confidentiality of the exchange between keypad and card. Some system designers are considering moving from the present handsets, which have very large numbers of keys, to a simpler cursor control device (more similar to a game controller), which may be easier and more instinctive to use, or even to a touchscreen. Either of these options would require a reconsideration of the PIN entry rules.

Systems must be capable of taking into account local contract laws and rules governing gambling and other services offered. An aircraft is within the jurisdiction of a country it is flying over, and so the rules may change during a flight.

The computer systems required to support such a system are potentially quite complex and costly. The designer must balance the needs of performance and reliability against cost, weight, and the likelihood that the system will be upgraded during its lifetime. As a result, systems are usually highly distributed (see Figure 16.2), and support for card payments (including smart cards) is handled in a separately defined subsystem.

Road tolling

Traffic growth is an acknowledged problem in many countries, particularly in large cities. With rising incomes and a higher demand for mobility, there seems to be no way to control the growth other than road tolls.

The technology that works best for road tolling is radio frequency identification (RFID). Like a contactless smart card, RFID sensors pick up radio signals from fixed antennas and use these signals both to power and to communicate with the fixed system. The systems vary considerably in power and the amount of data they can store or transmit, but those most attractive for road tolling allow a vehicle to pass at normal speed over an antenna embedded in the road. (See Figure 16.3.)

Most such systems would require the sensor to be fixed permanently to the vehicle. Until all vehicles are fitted with such a unit, this involves a

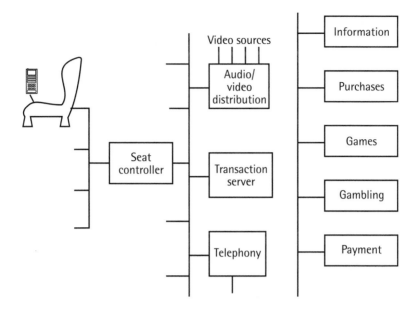

Figure 16.2 Inflight entertainment system.

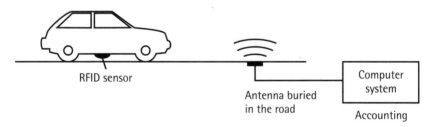

Figure 16.3 Road tolling using RFID.

significant extra cost. A way around the problem, at least in the shorter term, is to use a contactless smart card, which can be passed by the driver past an antenna alongside the car. This method requires the driver to slow down to 4 mph or less (there is usually a barrier, but an alternative is to photograph the license plate of cars that pass without paying), but it requires no additional equipment in the car and can be implemented alongside cash-based systems.

In Singapore, access to the city center is controlled by sensors mounted on overhead gantries. The sensors communicate by radio with boxes located on the dashboard or windscreen of the vehicle, which carry a CashCard electronic purse. The unit deducts a variable amount from the purse according to the time of day. If the vehicle does not have a working box, or if there are insufficient funds in the purse, a photograph of the number-plate is taken and the driver is fined.

Parking

Also related to traffic growth is the difficulty of controlling parking. Many local authorities have problems not only controlling illegal parking, but also collecting the fees due for legal parking. Parking meters and *pay-and-display* stations that collect cash are frequently vandalized. Local authorities want to be able to apply priorities to parking, giving preference to park-and-ride users and encouraging people to avoid peak times.

Such facilities are most easily provided by pay-and-display stations covering a parking lot or street and operated by cards. In some areas, smart cards are being used to avoid the risk of counterfeit or alteration of magnetic-stripe cards. Smart cards allow different groups of customers (staff or local traders, for example) to buy parking at different rates from the general public. Such facilities may only be available during certain hours or subject to restrictions on the length of time or frequency of visits. They may be implemented as part of a more general town card scheme, offering benefits to residents.

An alternative is to use a smart card in conjunction with a small device incorporating a real-time clock—the personal parking meter. This pro-prietary device can be placed within sight inside the vehicle, so security is less of an issue. Less "street furniture" is required, and there are no over-heads for the installation or maintenance of meters or pay stations. But a very high temperature specification is required for such a device! Since users must buy their own device, this is most suitable for areas where parking is both restricted and paid for.

As with the public transportation system, decisions must be made about the quantity of data to store and collect. Most pay stations operate off-line, with a periodic connection to the central system; parking meters are fully off-line, unless they use a low data-rate connection along the power system. In this case, it is rare to store any transaction details on the card itself, and even the terminal is more likely to store summaries than the details of each transaction.

There are few security implications for a card used only for parking—it is very similar to a telephone card. Where an open electronic purse or multiapplication card is used, the problems lie more with the card than with the application. These issues are dealt with in Chapter 18.

Where bank-issued cards are used to pay for parking, operators need to ensure that parking-lot and similar low-value transactions cannot be outlets for fraud. Fraudsters sometime use low-value unattended transactions to test whether cards have been reported stolen. Such terminals must have access to a comprehensive hotlist and be able to capture or block the card in the event of a hit.

References

[1]   Crotch-Harvey, T., "Operators Are Poised to Reap Smart Card Benefits," *International Railway Journal*, February 2000.

[2]   *Specification for Airline Industry Integrated Circuit Card – Version 02.* Resolution 791, Passenger Services Conference Resolutions Manual, International Air Transport Association.

17

Personal Identification

Identity card requirements

Although the function may be unique, an identity card can be used in many different ways. In an on-line system, the application itself can be performed by the central system, so the card may often appear to be a multiapplication card. For example, a company identity card could be used for access control, controlling and recording access to photocopiers and fax machines, paying for meals in the canteen, and using the company library.

For open systems, there must usually be two levels of identification: The card must be able to authenticate the issuing organization as well as the cardholder. This implies at least one register of unique organization codes and one common key system to perform the identification.

Thus, such systems can only work with open standards and a centrally controlled register of issuers. Increasing numbers of sectors are building these registers and public key infrastructures based on them.

Other organizations designing identity card systems usually depend on some proprietary features or arbitrarily chosen codes to ensure uniqueness. Designers would be advised, however, to avoid the situation that now exists for magnetic-stripe access-control cards, where many manufacturers have simply issued card numbers sequentially starting at 00001!

Issues

Level of security

A key decision for an identity card system is whether the card must:

+ Confirm the identity of the cardholder before disclosing any data;

+ Offer data for confirmation by an external device, system, or person;

+ Provide data to the system without any check as to who is using the card.

These correspond to different *levels of security*, and the answer will depend both on the type of threat envisaged and on the extent to which the card must be used off-line. The method used for *confirming the cardholder identity* will depend on whether it is being used in a face-to-face or unattended environment.

On-line and off-line systems

The extent of the system, both in terms of geography and applications, will determine whether or not it is realistic to have all checking carried out *on-line*. For a school or club building, for example, it is probably possible to have every door connected to a central system. For a national student card system, an on-line system could not be considered. Even where it is possible, on-line checking may be an expensive option; the ability of smart cards to operate off-line will often offset their extra cost compared with other systems.

There is often scope for compromise: an off-line system with exception files (such as hotlists) and key files (for new issuers) updated on a regular basis by modem or by broadcast.

The level of off-line use will govern how many data need to be stored on the card. Although smart cards can store several kilobytes of data, they are no match for the hundreds of gigabytes that can be stored on a central system. In most cases, the data on the card must be duplicated elsewhere. Holding the data on the card is, however, valuable if it must be used by many different systems.

For many ID cards, a memory card will be adequate; the card can still be authenticated using a static data-authentication technique, and the cardholder data, including any biometric template, can be stored in data files inside the card. Only where there is a serious threat of counterfeit cards, or where the card must also perform other functions, does it need to be an active device.

Card issuer responsibilities

The *role of the card issuer* is important in any identity card system, particularly for open or semi-open systems. In any system offered to the public, we have to look at the responsibility and liability the issuer may assume. Could it be held liable if information is disclosed or if a person gains wrongful access to a building? Is the card issuer able to control all aspects of the system or are some reading terminals controlled by others?

Access control

Alternative technologies

Almost every factory, school, and office building seem to need some form of access control nowadays. Smart cards face competition from other types of systems in this application, including biometrics using a central database, bar code cards, and magnetic-stripe cards, including secure magnetic-stripe cards (such as watermark cards).

Smart cards are more expensive than most other forms of cards, but their advantage lies in their ability to operate off-line and to perform other functions as well as access control. Many access-control schemes require only very small quantities of cards; this makes printing and personalization very expensive.

Features

A system must be able to handle visitors; either they must be registered into the central system or they should be issued with temporary cards.

Many building access systems combine different forms of checks. During normal hours access may be by card only, a card with a fixed code may be used at night and on weekends, and the same card with a PIN or biometric may be used for a restricted area. All controls are more effective if they are carried out in front of an operator or security person; in these cases an on-line system or the card itself can also display a photograph, the person's name, or other characteristics for the security person.

Many cards also have visual identification by means of a photograph or signature on the card; this is not always a good idea, as a person finding the card then knows whom to impersonate.

Other features required by many security systems include:

- Checks that one person cannot hold the door open for another (systems may count the number of people passing and record if more pass than were recorded);

- Checks against a card being passed back to another person waiting to enter;

- Complete records of entries and exits;

- Control of access by groups of cardholders according to the time, presence of a supervisor, or day of the week (including holidays). This type of control nearly always requires an on-line system.

Special cases

Card-based keys for hotel door locks may operate on-line or off-line. On-line locks can be reprogrammed with a new key every time and reset when the guest checks out. Off-line locks can be reset by an operator, for example when the room is made up at the end of a stay. Simple systems can be operated by magnetic-stripe cards, but more complex schemes including staff cards and access to other hotel facilities are more likely to demand smart cards. The lock can record which key was used to enter and when.

At the Olympic Games in Atlanta in 1996, athletes, team members, and staff were issued with contactless smart cards containing their identification number, height, weight, country, access rights, and a hand geometry template. At the entrance to all restricted areas, the hand geometry was read at the same time as the details from the card; if the card was authenticated satisfactorily, the biometric check passed, and the access rights were valid, then entrance was allowed.

Club membership cards often allow access as well as performing other functions. Where access control is the main function, a memory card with authentication functions is the most suitable type of card. If other functions such as cashless payment are to be included, the memory card may still be adequate if the system can operate fully on-line. Otherwise, a microprocessor card should be used.

Universities and schools

Universities and schools have been popular with smart card vendors as test sites; it turns out that the technology offers some real benefits in this sector. In most cases, the card functions only as an ID card; the use of smart cards makes the cards themselves more secure through the use of authentication.

Increasingly, however, the ID card is being combined with a cashless payment function. The card can be used in canteens, bars, and vending machines, and it can be used to pay for photocopying and other services on campus. It can also record attendance at lectures off-site and keep track of course credits; storing these on the card allows the student to check his or her records using a kiosk or off-line PC system.

In many British schools, pupils use their smart cards to record their attendance, to pay for school meals and vending machine snacks, and to gain access to computer rooms and other restricted facilities. The card allows pupils eligible for school meals allowance to receive them through the card, without other pupils being aware who is receiving free meals. In some instances healthy eating is also promoted by awarding points according to the type of food being bought; these are stored on the card, and the school can decide its own reward system.

Government cards

Smart cards could also be used as passports or national identity cards; realistically they are more likely to be used to supplement a printed card for regular travelers or other specific functions. The European Union has asked member states to make provision for a chip on any new driver's license cards; the obstacles to introducing such cards are more political than technical, although several standardization issues would also have to be resolved.

The national smart card project in Malaysia took personal identification onto a new plane. This involved plans to issue a smart card to every citizen. The card would have on it not only a national ID application, to be used by the cardholder in dealings with government bodies, but also a health application, an electronic purse, and an X.509 digital certificate for use in e-commerce transactions. This project stalled following the economic downturn in the region in 1998–1999, but may now be revived. The Singapore government has similar plans.

The South African government is currently seeking to implement a National Electronic Identification Card, which will include applications for an electronic passport (including fingerprint and color photograph), employment records, health records and prescriptions, housing, welfare, driver's license, vehicle ownership, firearms license, and an electronic purse.

Several other governments have announced plans for citizens to be able to carry out most of their interactions with government—applying for driver's and television licenses, registering births and deaths, filing tax returns—on-line. Many of these plans mention the possibility of chip cards but—at least in countries where civil liberties are high on the agenda, as in the United Kingdom—there is some reticence about making such a card compulsory. Nevertheless it is difficult to see how these types of transaction could be carried out securely without the use of a card or similar token. Perhaps a card with a different form factor would be seen as less threatening.

"White cards"

Although the underlying function of most cards is to identify the cardholder to a computer system, the idea that such cards could be truly personal is only now starting to come into perspective—it is not yet a reality.

Most cards are associated with a specific function, such as those we have described in the last few chapters. So we need an employee card for work-related functions, a bank card for payment and banking, a health card for insurance and medical data, and so on. Only in a very few instances are these functions combined.

Not everyone wants a single card for all purposes, indeed not everyone wants an ID card at all. Yet several governments have expressed the view that only a national government should certify a person's identity. If other organizations become involved in this, they are in effect issuing pseudo-identity cards.

One way around this is not to have any master organization certifying the cardholder as such, but for each application owner to issue certificates relevant to its application. This works best with the "white card," which we buy and load with the applications we ourselves want. Instead of one size fitting all, I can create a card—or several cards—which contain the applications relevant to my lifestyle. The application owner need not certify who I am, only assert my right to perform certain operations within its domain.

This is the most complex form of multiapplication card, which we will come to in the next chapter.

Many of the more futuristic scenarios dreamed up by technical journalists and technology enthusiasts involve personal preference cards. Drivers of Mercedes S-class cars have a chip card within their key which records each user's driving position, preferred air temperature, and other controls. There is a European standard for recording special needs onto cards [1], so that those with impaired sight can be presented with a larger font or brighter screen; others may simply need more time to enter a password than the norm.

Why should I not also have a card that records the way I like my coffee, what type of background music I prefer, and so on? Here the obstacles are not technical or political but organizational: Who would issue such a card and set the standards for recording preferences? It is likely that such a scheme will take off first within one specific sector, where a suitable coordinating body already exists and will spread from there by commercial alliances and cobranding.

Such a card must have the ability to record data under many different headings, probably in a very wide but shallow file structure. Because cardholders have different preferences for confidentiality and convenience, access to each field must be controlled by both the application and the cardholder. This could be handled within the existing ISO 7816 file structure, but it would not be efficient; a further generation of file structures may be required for this dream to be realized.

Reference

[1] EN 1332-4 (1999), *Identification Card Systems: Man-Machine Interface, Coding of User Requirements for People with Special Needs.*

18

Commercial Structures for Multiapplication Cards

Throughout the history of the smart card, writers have assumed that a single card will be able to perform many functions, across the whole range of applications discussed in the last few chapters. Although the technology for this is now available, few such cards exist in practice; this chapter explores the obstacles and the means by which they may be overcome.

Functions and applications

As we saw in Chapter 9, many cards can support what seem to be multiple applications (such as access control, vending, library lending, and photocopier control) without needing to have more than one application (program) loaded on the card.

In the case of a microprocessor card, the card would have one *application* loaded, but this application would include several *commands*: for example, authenticate card, authenticate terminal, send ID, increase purse value, decrease purse value. The application would, however, be developed and tested as a single unit, often as an integral part of the card's mask.

A memory card could be made to perform the same functions, but in this case the commands are executed by the terminal rather than by the card; this implies a higher level of control of the terminals where the cards will be used.

For an example, our company, Great Universal Machines, uses Supersoup Limited to operate the canteen, and bus services to and from the factory are provided by Buggins Bus Company. Great Universal decides to issue smart cards to all employees; these will be used for registering time and attendance, for access control out of hours and to controlled areas, and for logging on to the computer network. To give better control of the subsidized canteen and bus services, each employee receives a canteen subsidy each month, and money can also be loaded onto the card by a voluntary deduction from the salary or at a card-loading station. The bus pass is also revalidated at the beginning of each month (see Figure 18.1).

All of this can be handled by a single application on the card, yet still give adequate control to Supersoup and Buggins Bus (see Figure 18.2).

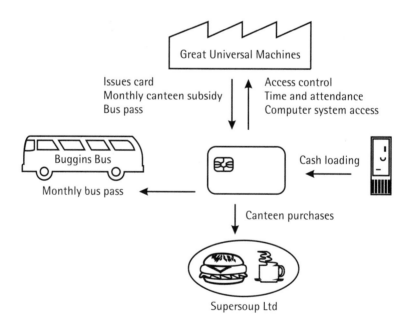

Figure 18.1 Great Universal company card.

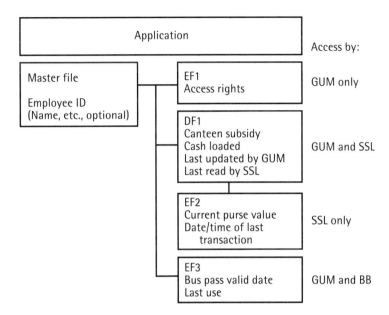

Figure 18.2 Card structure for Great Universal company card.

So when is a multiapplication card really needed? When several companies are responsible for the different applications, when the applications must be developed independently, and particularly when we may need to download new applications onto the card after it is issued. In this situation we need to protect each application and its data from the others. Chapter 9 covered the technical requirements and the products which have been developed to meet those requirements.

We could also envisage situations within a single organization where the implementation of a smart card program is phased; funding may not be available for all applications from the start, or some applications may simply have a longer development timescale. In these cases, the requirement for independence is more technical than commercial.

Downloading

Cards that can run multiple applications offer the opportunity to load further applications onto the card after it is issued. This allows applications to be updated or additional services to be offered to cardholders.

It would also allow the so-called "white card" scenario described in the previous chapter, in which I purchase a card (in much the same way as I would a PC) and then ask my preferred service providers (my bank, one or more transport companies, a telephone company, or my employer) to load *their* applications onto *my* card.

Although, as we saw in Chapter 9, today's multiapplication operating systems would allow this technically, they all assume a commercial structure in which there is a single card issuer who retains responsibility for the card throughout its life. It is the card issuer that decides which applications may be loaded onto the card.

Where other application owners or developers are allowed to download applications onto the card, it is likely to be restricted to a small circle of application issuers in the context of a cobranding or cross-selling agreement. Examples would be public transportation companies in adjacent areas, or a bank, insurance company, and stockbroker.

In this environment, the card must authenticate both the application issuer and the downloading station before it proceeds to download any application or data. It may well be necessary to carry out the whole operation in a secure environment or on-line to a trusted system. To authenticate the application issuer off-line, the card must carry the public key of a certification authority (which might be its own issuer) so that it can verify the certificate offered by the application issuer. Multos additionally encrypts the whole application and places it into a secure "envelope" signed by the card issuer and certified by Maosco.

The terminal should also be authenticated to avoid any danger of a subsequent replay; again, Multos has the additional security of a certificate, which is only valid for loading one application onto one specific card.

The risk of viruses affecting downloadable smart cards cannot be ignored. The issuer controls built in to the current operating systems form a first level of protection, but as with all software it is impossible for a card issuer to prove that an application contains no rogue code. The classic Trojan Horse effects of any such code should, however, be minimized by the interprocess protection—the virus-ridden application can only access its own data—and since cards never talk to each other directly there is no real likelihood of a virus spreading uncontrollably. There is, however, a real danger that any attempt to load an unauthorized application, even if it is not permitted by the card, might disrupt the normal operation of the card or cause inconvenience.

Hybrid card types

Several multiapplication combinations involve a public transportation or access-control application combined with a financial application. As we have seen, issuers in the first group have a strong preference for contactless cards, while those in the second are committed to using contact cards.

Where *combi* cards are used to overcome this problem, we must also consider whether it is appropriate for the applications to share memory—and, if so, whether the access control to each section of memory should be the same. Unless there is a need for the two applications to communicate, the answer is probably no: They should be kept separate. If this is not done, the security of the card, and hence of the system, suffers from a least-common-denominator effect: Each aspect of security is only as strong as the weaker of the two systems.

Special interprocess messaging must be set up to allow applications in these classes to share data. And it is often necessary to rewrite applications completely, taking into account the different characteristics of the two interfaces, not only in speed but also in the user interface. For example, in the contact card case, we can assume that users always know when their card is in communication with a terminal; PIN protection may be used for security, and abnormal termination of transactions is very unusual. For contactless cards, none of these is true: Transactions are often interrupted as the user moves into and out of the field, or as multiple cards contend with one another. Combi cards often have a reduced-power mode, whereby some functions (typically cryptoprocessing) are not powered up unless there is adequate power available.

Many current contactless card applications were developed for ASICs with dedicated encryption functions; microprocessor-based combi cards often do not meet the speed requirements of these applications without major redesign.

Card and application control

Where applications are updated, it is important to track (within the issuer's card-management system) which applications and versions are held on each card; as with all software, version mismatches can cause problems. Keys are often updated along with applications, and if an old set of keys is in use, this may compromise the security of other, newer applications on the card.

Many of today's card management systems are based on flat files. In a multiapplication environment, we have a complex relationship between cardholders, cards, application issuers, account numbers, application version numbers, keys, and certificates. It is found that a relational database is almost essential for managing this structure.

The most common approach today, when the use of multiple applications is limited and version updates infrequent, is to load all of the relevant applications onto every card and to enable and disable them when required. This allows all applications to be kept in step and only requires one set of flags in the card issuer system.

To meet the long-term need for multiapplication card control, a further generation of application management systems is evolving. At the time of writing, the most advanced of these are designed for the Open Platform—they include platform seven's Platform Management Architecture (PMA), which allows multiple applications to be managed across a large multiapplication card base. Unlike traditional card-management systems, these are based around the application rather than the card.

Issuer responsibilities

One factor that will deter many card issuers from opening up their cards to other applications is the fiscal and contractual responsibility that the card issuer may take on. This will partly depend on the law in each country; for example:

- In Denmark, there is a specific law for payment cards but not for other cards.

- In Germany, the card issuer is deemed to have a contract with the cardholder by virtue of issuing the card, but this would not apply to other application issuers.

- In the United Kingdom, the card issuer's responsibility includes liability for the goods purchased in the case of credit cards, but not for debit or purse cards.

In general, the card issuer will have some duty to ensure that the cardholder is not exposed to financial loss or exposure of confidential data as a result of normal use of the card. It will also have a responsibility towards those accepting the card (in payment or as proof of identity) and towards other application issuers.

If, for example, a card carries the name and account number of the cardholder, does this imply that any form of credit check has been carried out or is any guarantee of payment made or implied? Card acceptors in the retail and service industries are used to the terms of the financial card schemes, which do guarantee payment for validly performed transactions.

Issuers of multiapplication cards must look carefully at the legal liability they could be accepting and ensure that the cards and mechanisms they propose will provide adequate protection.

Consumer issues

Cardholders who receive a card by virtue of their employment, as benefit claimants, or as pupils at a school are entitled to protection of their personal data in the same way as other citizens, but are otherwise unlikely to have any say in how the card is used. Card issuers often seek to include a blanket permission clause in their application forms; by signing the form consumers give the issuer the right to use all the data for the card issuer's business purpose, but under current data protection rules, in Europe at least, that right would almost certainly not extend to other application issuers or their business partners, unless the cardholder has given further permission.

Where the cardholder has chosen to have a card (whether it is issued free, as part of a service, or directly purchased), he or she has the right to know what is stored on the card and how it may be used. In the case of a multifunction card, it may be difficult to answer either of these questions, particularly if no one organization has full control of all of the data. Usually the card issuer retains ownership of the card itself; for a multifunction card this may no longer be appropriate.

There is a special case in which a government decides to issue a national identity card, driver's license, or other mandatory card. A general multiapplication capability in this case is almost certainly taboo in most western countries (although it would be the ideal medium for a totalitarian state). It is essential to define the data and its use before the card is issued.

Interchange and compatibility with existing card systems

A further set of problems arises when a multiapplication card is introduced to an environment that had been single application. Assumptions made about keys and certificates will often no longer be valid, and both card and terminal must carry the additional data needed to identify application issuers as well as card issuers.

Although the latest generation of card-operating systems takes us much closer to a full multiapplication environment, the migration of applications will take many years.

19

Designing for Security

Aims

The reader will by now be aware that there is no such thing as perfect security. A security design is a balance between the requirements of confidentiality, authentication, and integrity on the one hand, and convenience, cost, and reliability on the other (see Figure 19.1). The only way to ensure that no one can ever read your confidential data is to destroy it completely!

The more realistic aim of every security system should be to ensure that *the effort required to break the system exceeds the reward.*

Reducing the reward

We reduce the severity of the risk by:

- *Limiting the part of the system affected:* Systems should be subdivided so that no one key gives access to the whole system, and that loss of one piece of data only affects a limited part of the system.

- *Limiting the time for which the system is exposed:* Exposure can be greatly reduced by reducing the life of keys and other critical data.

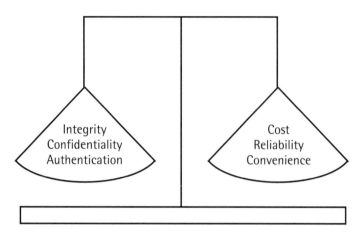

Figure 19.1 Security in the balance.

- *Limiting the amount at risk:* The value attributed to one card should be limited according to the strength of the security mechanisms used.

- *Reducing motivation:* Avoid setting the system up as a target by maintaining an atmosphere of trust combined with control and avoiding high-profile claims.

The last element is particularly important in deterring hacking attempts because for many hackers the intellectual challenge, and the publication of their achievements, is its own reward.

Increasing the effort

The effort required to break the system can be increased by all the methods discussed in this book.

- *Focus on the data at risk:* Do not increase the security of unimportant areas unnecessarily (unless as a deliberate decoy); it will distract attention from the more important areas.

- *Set security objectives:* These must be consistent and realistic. Do not demand military-level security unless you can afford it.

- *Choice of card type:* Select a card with a suitable level of security; specifications are rising all the time.

- *Encryption of data:* Use commercially available products, but ensure that they were designed for the purpose you are seeking.

- *Card, terminal, and system authentication:* These will be discussed later in this chapter.

- *Cardholder verification:* Use PINs, passwords, or biometrics.

- *Key management:* As soon as you start to use encryption, you also have to manage keys. Use commercially available systems where possible and follow the procedures recommended.

- *Control of access rights:* Access rights must be controlled both within the card and in any associated computer systems.

- *Physical security:* This remains important even where automatic checks are in place.

- *Organization:* Tight organizational control is essential to any secure system.

- *Procedures:* Procedures must be written and tested.

- *Audit and external verification:* The system should be independently reviewed and monitored.

Criteria

It will by now be clear that smart card system security is not a single-dimensional effect, and that each case must be looked at on its merits.

Nevertheless, as a starting point, we propose in this chapter a sequence of requirement setting, design, and analysis to assist in designing and assessing any system that seeks to use smart cards to enhance data security.

Types of security

The first step is to determine the broad requirements for:

- *Confidentiality:* What data must be kept confidential and from whom? What is the value of the information in the wrong hands?

- *Authentication:* How important is it to ensure that the card being used has been issued within our scheme? Must we also authenticate the person or is it the card that matters? Where the terminal can alter data on the card, must the card also authenticate the terminal?

◆ *Integrity:* What would be the effect if data were lost or altered during transmission and storage?

◆ *Nonrepudiation:* How important is it to ensure that neither party can dispute the transaction after the event?

Ideally, we should go on to set quantitative criteria for each of these areas as described in Chapter 3. However, we can use a simple model to construct the outline of a system; this model can then be analyzed and the appropriate strength of protection applied to each element.

Model

The simple model assumes that data must be stored, transmitted, and used. At each stage we must apply the appropriate level and type of security mechanism (see Figure 19.2).

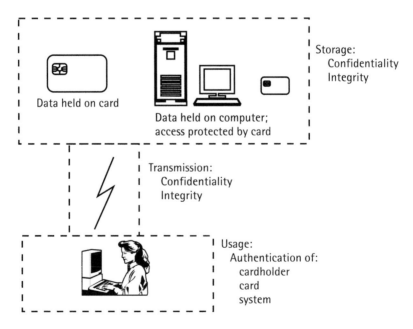

Figure 19.2 Security model.

Storage

We start with the requirement for data storage. If the data must be portable or accessed off-line, we can store it in a smart card; otherwise, it will be preferable to store the data on a computer system and use the card to gain access to that data.

According to the level of confidentiality required, we can:

* Store it in clear on a memory card or a computer file;

* Store it in a protected memory card or on a computer with access-control rights controlled by a password or card system;

* Encrypt the data and store it in a secure memory or microprocessor card.

Data is made unintelligible to unauthorized persons through encryption with secret keys, usually with symmetric algorithms. We can increase the strength of the encryption by increasing the length of the key.

We must also check that data stored is not altered, and that data cannot be lost as a result of a malfunction, misoperation, or power failure. Again, there are several levels depending on the degree of integrity required.

* At the lowest level, data in the card is assumed not to change except during a controlled operation. A checksum may be held for computer files. Daily backups are made from all central files, and transaction logs are kept for several days.

* For added security, checksums can be associated with all data. Every data item that changes is logged; transaction logs within the card are usually quite small (5 to 10 transactions) but on a central computer system they may be kept for weeks.

* For demanding applications, MACs are associated with critical fields. Important data is also held in shadow files on the central system.

Transmission

When data are transmitted from one system to another, we must ensure that the information is not altered, accidentally or deliberately, on the way. If it is confidential, then we must protect the interface either physically or by encryption. We check the integrity of messages by using cyclic redundancy checks (CRCs), transaction counters, and MACs:

* A message CRC is the most basic check; it must be long enough to pick up any accidental change, but it provides no protection against deliberate alteration.

* A MAC that includes the message number, date, time, and key data elements—or a digest of these elements—provides good protection against alteration. (Message numbering is very easy to implement and should be part of any system, no matter how low the security requirement.)

* To include checks for nondelivery, not only the data but also the acknowledgment should be MACed.

Use

When the data are used, we must perform checks to ensure that the person presenting the card is the cardholder. It may also be necessary to check the authenticity of the card and of any terminal or system in which it is being used.

* For many systems, possession of the card is sufficient to give access or to allow its use. Cards carrying a signature or photograph can be checked manually in face-to-face environments.

* The simplest form of automatic cardholder authentication is by a PIN. Although this method has limitations, its ease of implementation, customer acceptance, and proven record weigh strongly in its favor.

* For higher security applications, several effective biometric checks can now be implemented through smart cards.

When it comes to authenticating the card itself, there are again several options:

* The simplest method is to check the card identification held in PROM. This is normally only suitable for closed systems.

* For open systems, we must store an individual cryptographically derived checksum in the card, or resort to public key cryptography. A simple asymmetric scheme does not require the card to perform public key encryption.

- Where the card can perform public key encryption fast enough, we can use two-way dynamic authentication. This is the most secure way of authenticating a card.

Finally, we must consider the need to authenticate the terminals and systems in which the card will be used.

- Again, there is a basic check in that the cardholder does not offer the card unless he or she believes that the terminal is authentic.

- The card may supplement this by checking a simple symmetric or asymmetric key certificate held in the terminal.

- For maximum security, two-way identification should be used; this is particularly important when operations that change data in the card may be performed off-line.

For large-scale open systems we must look to a public key infrastructure. It will often be possible to use an existing PKI, although we must make sure that the commercial structure (as well as the technical structure) meets our needs. Otherwise we must implement a PKI in-house, by buying a software package and operating it ourselves, or by making use of a managed service.

The model is summarized in Table 19.1, where we list a range of typical tools that may be used to meet low, medium, and high requirements for each of these conditions.

Analysis

Initial situation analysis

Having ascertained the level of requirement in each area and the outline of the system to be used, the next thing to consider is the pattern of risk. Risk analysis was considered in more detail in Chapter 3. We must consider a very wide range of outcomes, both positive and negative, and assess for each the severity and likelihood of the outcome.

If there is any gain to be obtained by attacking the system, then the designer must assume that attacks will occur. He or she must consider the likelihood and possible severity of each of these attacks: How persistent or well motivated would the attacker be? What resources could they have

TABLE 19.1 Security Requirements and Tools

Parameter	LEVEL OF REQUIREMENT		
	Low	Medium	High
Confidentiality	Data stored in clear on memory card or computer	Data stored in protected memory card or on computer with access control	Data encrypted and stored in secure card
Authentication (cardholder)	None (possession of card is entitlement)	PIN	Biometric
Authentication (cards)	Check card identification in PROM	Individual cryptographically derived checksum held in card (SDA)	Two-way dynamic authentication (zero-knowledge protocol)
Authentication (terminals and systems)	None (cardholder does not offer card unless satisfied)	Symmetric or asymmetric key certificate held in terminal; consider use of PKI for open schemes	Two-way identification, both for off-line and on-line operation; some form of PKI likely
Integrity (communications)	Message CRC	MAC, to include message number, date, time, and key data	MAC as left, with acknowledgments
Integrity (storage)	Checksums for data held on computer files only; daily backups	Checksums associated with all data; full transaction logging	MACs associated with critical fields; important data also held in shadow files

access to? How much access would the attacker have? Would it depend on a range of coincidences or would the opportunity always be present?

Sources of attack

All possible sources of attack should be considered, for example:

- *Normal cardholders* will try to do things that the system is designed to stop them from doing or they may use the system in ways the designer has not considered. Often such attacks are not malicious or even intentional, but they are seen as attacks by the system.

- *Careless cardholders* often lose or damage cards or terminals. Often some of the card issuer's best or most important customers are careless, and customer-service requirements dictate that the system must be able to recover quickly from this situation to restore the cardholder's rights.

- *Malicious or untrustworthy cardholders* have the highest level of access to cards; they may gain access to systems or provide cards and codes for analysis.

- *Insiders*, typically employees of the card issuer or scheme operator, have opportunities to copy, analyze, or steal data and hardware, or to give special privileges or benefits to their friends (*sweethearting*).

- *Outsiders with system access*, for example maintenance engineers or telephone company employees, are often a particular threat as their activities are specialized and not fully understood by those supervising them.

- *Card thieves*, even if they have a card but no other information can often launch a limited attack. Or what if they also know the identity of the cardholder and possibly some additional information held in a diary or wallet?

- *Criminal gangs* may gain access to dozens or hundreds of cards in a systematic way; they will also have access to computer systems and have the time to set up bogus accounts or businesses.

- *A serious and sustained effort to break the system, for its own sake* is the most dangerous of all, but is only likely to be suffered by government or national financial systems. Political and religious fanatics are very dangerous adversaries because they do not stop at the boundaries set by effort and reward.

As well as determining how easily each of these possible attackers could gain access to vulnerable elements of the system, the designer should consider how easily that penetration could be detected.

Quantitative analysis

Where we are able to set quantitative criteria, in the form of MTBIs or a probability value for each type of incident, the design can now be tested against these criteria. This can be done most conveniently by setting them out against each other in a spreadsheet: For each risk we consider the possible causes and modes of failure, their likelihood, and the cost of the next level of countermeasure.

Armed with this information, we are able to state confidently not that the system is 100% secure, but that we have calculated the risks and assured ourselves that further countermeasures are not justified by the risk.

Risk analysis checklist

Table 19.2 gives a simple checklist that may be used to analyze risk in a smart card system. It is not comprehensive, and will need to be adapted for each situation, but it will give guidance to a designer seeking to analyze a system and to locate areas of high risk.

For each of these cases, a range of possible outcomes should be considered, and each should be assessed for severity and likelihood.

The BS7799 standard referred to in Chapter 3 will also act as a checklist for the system as a whole.

TABLE 19.2   Risk Analysis Checklist

IMPLEMENTATION	OPERATION
Design process	Hardware failures
Equipment and card selection	Software failures
Public key infrastructure	Card failures
Software design and selection	Card issuing
Fitness for purpose	Use (consider each instance)
Testing	Lost and stolen cards
Delivery timescales	Expired cards
Implementation timescales	Counterfeit cards
Project management	Breach of confidentiality—card data
Personnel	Breach of confidentiality—computer data
Training	Cardholder authentication: false accept/reject
	Card authentication: false accept/reject
	Terminal/system masquerade
	Key compromise
	Effects on other systems/customers

Further iterations

Having completed the process once, we need to compare the results of the analysis against our original requirements. If any of them is out of step we must go back and repeat the process.

Following any significant design change, the security design and implementation should be reviewed. It is also a good idea to repeat the whole process periodically—say, once every year or 18 months—because although the principles do not change, the tools available both to the hacker and the system designer are evolving continuously.

External audits are often recommended, and the security of any external system, such as a network or card personalization system, must be reviewed and if necessary tested.

20

Looking Forward

Market forecasts

The smart card market grew from around 500 million cards a year in 1996 to almost 2 billion in 2000. Europe currently accounts for around 65% of the market, with the balance almost equally split between Asia-Pacific and the Americas. The most rapid growth in the last year or two has been in Central and South America.

Several Asian countries have large smart card programs; Singapore, Hong Kong, Korea, and China all have major projects. These continue to have very strong growth potential—many projects were delayed as a result of the economic downturn in that part of the world in 1997–1998, and are now expected to come back into play.

North America continues to lag behind the rest of the world; although there are many small projects, none has yet reached the level of true mass-market rollout. However, the need for Internet authentication described in Chapter 13 alone is expected to drive demand in North America up to perhaps half European levels by 2005.

In the mid-1990s, when the smart card market was growing at around 100% a year, manufacturers and market research companies fell over each other to produce the most optimistic estimates. These days, the market is more mature and forecasts more conservative. Nevertheless, volumes are expected to rise to between 4 and 5 billion cards by 2005 (see Figure 20.1).

Telephone cards will still remain the largest application by volume, accounting for around half the total, but many of these will be microprocessor cards rather than the memory cards that have dominated this sector up to now. Transportation cards, most of which will be contactless, will rapidly become the next-largest sector. Despite the rollout of the EMV credit/debit cards described in Chapter 14, bank-issued payment cards are likely to be overtaken by PC authentication, although it is still not clear who will be the primary issuers of such cards. The higher value multifunction cards will start to have an impact in 2002 or 2003, particularly affecting the total value of shipments, since their unit value is around 20 times that of the simplest microprocessor cards.

There will be a growing demand for services surrounding the smart card industry, which up to now has been dominated by card and chip manufacturers. There are already specialized software and security companies, certification and PKI service operators specializing in chip card systems, and more and more mainstream software and security companies have chip divisions. At the operational level, card issuing and personalization bureau, maintenance organizations, and vertical-market scheme operators are marketing their services across a wide range of countries and applications.

In some sectors, and particularly in the United States, the market for closed systems (in companies, schools, and universities), and for commercial services operating in a specific niche or region, is likely to grow faster than that for full-scale public systems. Closed systems are easier to manage,

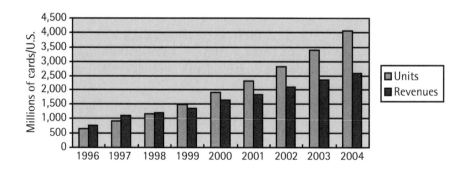

Figure 20.1 Forecast smart card market growth. (*Source:* Frost & Sullivan.)

and it is easier to pinpoint and control potential points of weakness. However, many closed-system designers also incorporate proprietary features that appear to improve security. As we have seen, this is not always a safe option, as the level of care and expenditure that goes into the design of the proprietary feature is likely to be less than for the corresponding commercial product.

Cross-sector schemes are still seen as being difficult to manage and their growth is likely to be slower. Government-sponsored schemes have a long gestation time, but there are now enough such projects in the planning phase that one or two must come to fruition by 2005. If these are successful, other governments are likely to follow, and this could result in a major change in the structure of card systems and the way they are viewed by the public.

Cards

Chips

Memory cards are now a relatively stable product, and there is little demand for upgrading in this area, other than gradual increases in memory size. More and more applications that would have used memory cards a few years ago are now turning to microprocessor cards, which offer more features at little extra cost. Perhaps importantly too, the terminals that accept cards are increasingly designed to handle the microprocessor card protocols, and are often seeking certification to the EMV bank card standards.

Current smart card chip fabricators are working at 0.5μ and 0.35μ, with supply voltages of 5V and 3V. Designs are already available for 0.25μ and 2.5V, but volume production has not started on these products. Although the semiconductor manufacturers have lines working at 0.2μ and smaller, they are unlikely to allocate these to relatively low-volume chip card products when there is still healthy demand for high-volume memory chips with equally good margins.

Japan's MITI has forecast that by 2005, 0.07μ, 0.75V processes will be available, moving on to 0.035μ and 0.45V by 2009. Smart card chips are likely to lag this by one or two years, and so are likely to be using 0.1μ, 1V processes in 2005.

Researchers are now considering the limits to this process: As the semiconducting layer becomes thinner and the individual features narrower, we start to reach the level in which the number of free electrons becomes a significant factor. Current and voltage can no longer be regarded as continuous but rather the movement of discrete units of charge.

Conventional wisdom still suggests that the maximum size for a smart card chip is 20–25 mm^2. Some manufacturers are breaching these limits, particularly for contactless cards, and it seems that for certain applications cards with these larger chips are reliable.

Most current high-volume smart card microprocessors are 8-bit designs. High-end devices, particularly cryptocontrollers, have moved on to 32-bit designs and there are some 16-bit processors. RISC and Java-optimized devices have already made an appearance, and are likely to feature in most manufacturers' portfolios within a year or two.

FRAM memories are likely to appear in volume production by 2005, and will largely replace E^{2}PROM within a few years. FRAM will enable much larger memory sizes: Instead of the 3–8 KB now common, 512 KB or more can be fitted onto a standard smart card chip at 0.1μ.

From a security point of view, chip designs become ever more complex, and more of the tricks hitherto applied only to top-of-the-range chips are now being used in standard designs. More chips incorporate detection for out-of-specification operation, including temperature and rate-of-change detection as well as voltage and clock speed. On-board clock generation is now common, to prevent timing-driven attacks; internal clock speeds up to 50 MHz are now possible and higher speeds are in view, particularly for RISC and pipeline processors. Processes such as the addition of a silicon insulating layer are now available, to protect against ion beam and electron microscope attacks. And most security-conscious chips have specific features to protect against differential power analysis.

Crypto-coprocessors

The performance of cryptoprocessors, and the degree of optimization for common chip card application functions, has increased dramatically over the last few years. Chips now entering production offer DES encryption and decryption in around 25μs, with 1,024-bit RSA signature generation in 100–150 ms. This is acceptable for all but the most demanding public transportation applications. Even standard 8-bit chips have acceptable DES performance.

Cryptoprocessors now encompass a wide range of symmetric and public-key algorithms. Many now have on-board public-key-pair generation, random number generation, and result verification functions.

Masks

The role of the mask is changing; where up to now the design of the "hard mask" has been a key part of the card manufacturer's skill set, and one of its major differentiators, the proprietary design is now regarded as more of a liability than an asset, and the aim of the mask designer is to implement a standard specification or interface.

Specifications such as JavaCard, Windows for Smart Cards, and EMV are superseding not only "vanilla" flavors of ISO 7816, but also the proprietary functions designed by each card manufacturer for specific vertical-market applications. Even crypto functions are designed to meet international standards and the JavaCard APIs.

Some interprocess protection features can now be taken for granted on all but the simplest cards. More advanced cards are normally designed to work with one of the three main multiapplication operating systems: Multos, JavaCard, or Windows for Smart Cards.

Contact/contactless

Although contactless cards are still held back by the proprietary nature of the specifications, and indeed of many applications, the major players in this field have shown that it is possible to create stable systems with excellent performance and few apparent security problems. While banks and security institutions still express some concern over the security of the radio interface, the transportation industry is moving ahead and implementing solutions on a large scale.

There are now combi versions of several popular cards available, and we expect this trend to continue; as in many other fields it is possible to separate the card communication method and protocols from the applications within the card, and so both sets of developments can continue in parallel.

Application downloading

Although all the technology is in place for application downloading, the commercial structures and download location infrastructures have yet to be developed. It is likely that closed environments such as universities, schools, and corporate campuses will lead the way in this field, just because one organization has most of the factors under its own control. This will be followed by high added-value applications, perhaps combining transportation, travel, and entertainment functions.

The field remains open for an enterprising organization to issue cards and set up an application-loading infrastructure, and then to seek customers in the transport, banking, municipal services, and health sectors. This possibility was explored in Chapter 18.

Card and terminal standards

Cards

The main parts of ISO 7816 are now universally accepted, and ISO 14443 for contactless cards is reaching the same status. The development of the further sections of these standards (such as the security architecture and commands) has been very slow and may be made irrelevant by the multi-application card developments. These are now becoming important standards in their own right.

At the sector level, there has been a dramatic increase in international cooperation and international standards-making in the smart card field over the last few years. Indeed, it is now arguable that the problem is less one of a shortage of standards than of a surfeit.

The development of standards such as GSM, EMV, CEPS, PC/SC, OCF, ITSO, and IATA 791 represent an opportunity for manufacturers to produce products on an economic scale and give stability to systems designers. The absence of such standards had previously held back implementations in their sectors. Perversely, however, these standards may now become less important: Once it has been shown how to achieve interoperability, it may be in the interests of users for companies to develop their products beyond the limited scope of the standards.

They must be able to take into account national institutions and legislation, local customs, and the individual needs of consumers. Where standards tend to make products homogeneous, much of the attraction of the chip card lies in its ability to be customized and to meet individual, personal requirements.

Terminals

In sectors where there will be many card issuers or schemes (as in banking, car parking, or airline ticketing), card acceptors will insist on one single terminal accepting all cards. Although the ISO 7816 standards do not guarantee this level of interoperability, most sectors do now have their own common standards. Some of these standards suffer from the problem of specification-by-committee: They offer a lowest common denominator solution, which does not always take into account the need for future upgrades and additional functions.

As with the cards, several companies have proposed interpreter-based terminals (such as Europay's Open Terminal architecture for financial terminals). However, others have felt that this architecture is too restrictive, and have sought to divide the implementation into several layers, so that an upgrade would in general only affect one layer.

Chip cards and the mainstream

For many years, the use of proprietary standards and interfaces kept smart card developments in an arcane world out of the mainstream of computing. The advent of JavaCard and Windows for Smart Cards is set to change this, although the transformation is by no means complete. The use of standard languages, development kits which are widely distributed, and the ability to share applications across platforms mean that many more programmers have access to the technology.

At least as important, their managers can see commercial benefits in working with what is now an internationally recognized route to security, portability, and cross-sector applicability. The close association of chip cards with such developments as e-commerce and m-commerce, on-line banking, and electronic ticketing makes them an appealing target for innovative development organizations.

Industry associations such as Global Platform, OpenCard Forum, the Smart Card Forum in the United States, and Smart Card Club in the United Kingdom, together with other similar groupings in other sectors and countries, have helped to increase awareness of chip cards and to promote their use. Looking forward, it will be important for chip cards to be seen less as a technology in their own right and more as an integral part of the computing, communications, and security sectors.

Instead of the rather rigid structure that has existed in the smart card business up to now (see Figure 20.2), many more businesses in the computing and application development mainstream now have chip card divisions, or chip card skills within other departments. The capabilities of the card are now recognized, much of the early hype and problems have been discounted and the chip card is becoming a normal business tool.

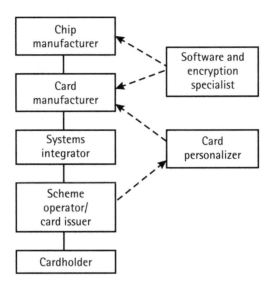

Figure 20.2 Smart card industry structure.

Conclusions

As with governments, presidents, and marriages, technologies often go through a sequence of wariness and courtship, honeymoon and disillusionment before we finally learn to live with them and to benefit from them despite their limitations.

Managing expectations is the key to this process: public awareness of smart cards has been affected by a large amount of news coverage; much of this is exaggerated, sometimes to the point of fantasy. Claims that smart cards are "absolutely secure" were always bound to lead to disappointment. Smart cards are small and flimsy; systems are set up and managed by people, and people are fallible. The more we protect our data using invisible, electronic means, the more important the physical controls, procedures, and checks become.

It has been comprehensively demonstrated that smart cards are, for many applications, the most economical and convenient way to store data and keys with an adequate level of security. As the technology develops, multiapplication cards and freely downloadable functions will become available, posing as many new problems as they solve.

Smart cards are here to stay, and will be used in increasing numbers of applications to provide security. Alongside the card itself sit many helper applications, technologies, and skills, which provide the security and make the card easier to use. This book has sought to make designers and managers of chip card systems aware of these tools, and to encourage their use in creating systems, which balance the needs of security and ease of use, cost and marketing potential.

The small size of chip cards belies their power and the number of skills which go into their creation. To use them effectively, the system designer must be psychologist and physicist, disciplined and creative. Chip cards are a powerful tool—and like most powerful tools, must be handled with care.

Appendix A:
Standards

In this appendix we list the most important international standards applicable to chip card systems and their security.

Standard Name	Title and Description
ANSI X3.92	Data Encryption Algorithm (DEA). The main source for the DES algorithm.
ANSI X9.19	Message authentication check. The most widespread form of symmetric-encryption MAC.
ANSI X9.30-2(1993)	Public key cryptography using irreversible algorithms for the financial services industry. Part 2: Secure Hash Algorithm (SHA-1).
ANSI X9.31-1	Public key cryptography using reversible algorithms for the financial services industry. Part 1: The RSA Signature Algorithm. RSA is by some margin the most widely used public key encryption algorithm. Although the algorithm is patented, RSA Data Security licenses it for general use and this standard gives recommendations in this respect.

STANDARD NAME	TITLE AND DESCRIPTION
CB	The French GIE Cartes Bancaires has defined a set of standards for use in France in connection with bank payment cards, all of which are chip based. The scheme operates entirely within France. The CB standards include the *Manuel de Paiement Electronique*, which is a functional specification, and the CBSA, CBST, and PCEMA standards, which define the protocols. A further standard defines the B0'mask used in French domestic bank cards.
CEPS	Common Electronic Purse Specifications: A set of business requirements, technical and commercial specifications covering interoperability of electronic purse cards. This standard is issued and maintained by CEPSCo LLC: www.cepsco.org.
EN 726	Identification card systems: Telecommunications-integrated circuit(s) cards and terminals. This standard was produced by the ETSI Working Group TE9; its seven parts cover all aspects of the design of a multifunction IC card with the exception of the application itself.
EN 1038	Identification card systems: Telecommunications applications, integrated circuit cards for payphones. This standard, which was produced by CEN, complements EN 726.
EN 1332	Identification card systems: Man-machine interface Part 1: General Design Principles Part 2: Card Orientation Part 3: Keypads Part 4: Coding of User Requirements Part 2 defines the "notch" to help blind people orient a card correctly. Part 3 defines the layout (telephone-style), tactile identifier, color, and arrangement for keypads on card-operated devices. Part 4 defines how details of a user's preferred interface can be stored on the card. These preferences could include large characters on the screen, speech prompts, more time for key entry, or amplification of sound output.
EN 1546	Identification card systems: Intersector electronic purse. A development from EN 726 and EN 1038, but with banking industry as well as telecommunications industry involvement.

Standard Name	Title and Description
"The EMV standards"	Europay, MasterCard, and Visa integrated circuit card specifications for payment systems. The EMV standards define the content, structure, and programming of chip-based payment cards. The electrical and protocol sections refer to any payment card, but the application and security sections apply only to credit and debit cards. They are designed to allow a single ATM or retail terminal to handle any bank-issued card, but without restricting unduly the ability of the card issuing bank to define how its payment scheme should work. The EMV specification is now owned and maintained by EMVCo (www.emvco.com). A revision (V4.0.0) was published at the end of 2000.
GSM	Global system for mobile communications: This set of standards covers the protocols for communication, the subscriber identity module (SIM), and other aspects of GSM operation. Some parts of this specification are only available to members of the GSM consortium. The SIM specification is GSM 11.11 (ETS 300 608).
IATA	The International Air Transport Association's Resolution 791 covers an electronic ticket to be held in a smart card.
ISO 7498-2	Open Systems Interconnection Reference Model: Security architecture. The OSI model is the framework for most modern data communications standards. Although some of the most widely used standards do not comply with OSI recommendations in all details—the TCP/IP protocols commonly used in wide area networking are not an OSI standard—the concept of a layered implementation, with specific services carried out at each layer, is still followed. The security architecture makes recommendations for the security services (from user authentication to non-repudiation of delivery) that should be provided at each layer.
ISO 7810 (1985)	Identification cards: Physical characteristics. This standard describes the shape, size, and environmental requirements of a plastic card to be used as an identification card. In practice, this standard size of card (85 × 54 mm) is used in almost all card applications.
ISO 7811	Identification cards: Recording technique. The six parts of this standard cover embossing and magnetic-stripe recording.

Standard Name	Title and Description
ISO 7816-1 (1998)	Identification cards: Integrated circuit cards with contacts. Part 1: Physical Characteristics. This covers all ISO 7810–sized chip cards with contacts, including memory cards and microprocessor cards. It includes specifications of the environmental and physical strength characteristics required.
ISO 7816-2 (1999)	Identification cards: Integrated circuit cards with contacts. Part 2: Dimensions and Location of Contacts. This standard defines the internationally accepted position for the chip (center left). The standard defines eight contacts, although most manufacturers only install five or six.
ISO 7816-3 (1997)	Identification cards: Integrated circuit cards with contacts. Part 3: Electronic Signals and Protocols. This key standard defines the way a chip card communicates with the outside world. It includes the synchronous protocols normally used by memory cards as well as the asynchronous protocols more often used by microprocessor cards. ISO 7816-3 defines the structure of the "answer to reset" with which a chip card powers up and initiates communication with the terminal. It also defines the way a card should respond to over- or under-voltage, and offers a protocol for negotiating a voltage and clock speed with the accepting device.
ISO 7816-4 (1995, Amendment 1 1997)	Identification cards: Integrated circuit cards with contacts. Part 4: Interindustry Commands for Interchange. This standard defines the content of several message types (including answer to reset). It also defines a file structure for use within a chip card, consisting of a master file (MF), dedicated files (DFs), and elementary files (EFs). The file and data object addressing provides a security architecture which is described in detail. It includes provision for applications to restrict access to whole EFs or to data objects, based on either authentication or encryption.
ISO 7816-5 (1994, Amendment 1, 1996)	Identification cards: Integrated circuit cards with contacts. Part 5: Numbering System and Registration Procedure for Application Identifiers. This standard appoints KTAS in Denmark as the international body able to designate national authorities, which in turn can issue and authenticate application IDs.
ISO 7816-6 (1996, Correction 1 1998)	Identification cards: Integrated circuit cards with contacts. Part 6: Interindustry Data Elements. This is a directory of data elements for use in ISO 7816-4 applications.

Standard Name	Title and Description
ISO 7816-7 (1999)	Identification cards: Integrated circuit cards with contacts. Part 7: Interindustry Commands for Structured Card Query Language (SCQL).
ISO 7816-8 (1999)	Identification cards: Integrated circuit cards with contacts. Part 8: Security related interindustry commands.
ISO 7816-10 (1999)	Identification cards: Integrated circuit cards with contacts. Part 10: Electronic Signals and Answer to Reset for Synchronous Cards.
ISO 8583 (1993)	Financial transaction card originated messages: Interchange message specifications. A revised version was due for publication in 2001.
ISO 8731	Banking: Approved algorithms for message authentication. Part 1: Data Encryption Algorithm. Part 2: Message Authenticator Algorithm.
ISO 9000–9003	Quality management and quality assurance standards. ISO 9001: Design, development, production, installation, and servicing. ISO 9002: Production, installation, and servicing. ISO 9003: Final inspection and test.
ISO 9564 (1991)	Banking: PIN management and security. The two parts of this standard define the techniques and approved algorithms recommended for handling PINs securely in a retail banking environment.
ISO 9796	Security techniques: Digital signature scheme giving message recovery.
ISO 9797	Data cryptographic techniques: Data integrity mechanism using a cryptographic check functions employing a block cipher algorithm.
ISO 9798	Security techniques: Entity authentication mechanisms.
ISO 9992	Financial transaction cards: Messages between the integrated circuit card and the card accepting device. Part 1: Concepts and Structures. Part 2: Functions, Messages (Commands and Responses), Data Elements, and Structures. These standards are not widely used.

STANDARD NAME	TITLE AND DESCRIPTION
ISO 10118	Information Security Techniques: Hash functions.
	Part 1: General Principles.
	Part 2: Hash Functions Using an N-bit Block Cipher Algorithm.
ISO 10126-1	Banking: Procedures for Message Encipherment (wholesale).
	Part 1: General Principles.
	Part 2: Data Encryption Algorithm (see also ANSI X3.92).
ISO 10202	Financial Transaction Cards: Security architecture of financial transaction systems using integrated circuit cards.
	Part 1: Card Life cycle.
	Part 2: Transaction Process.
	Part 3: Cryptographic Key Relationships.
	Part 4: Secure Application Modules.
	Part 5: Use of Algorithms.
	Part 6: Cardholder Verification.
	Part 7: Key Management.
	Part 8: General Principles and Overview.
	Part 6 is mainly concerned with PIN verification, although an annex allows for passwords or biometric methods.
ISO 10536	Identification Card: Contactless integrated circuit cards—close-coupled cards.
	Part 1: Physical Characteristics.
	Part 2: Dimensions and Location of Coupling Areas.
	Part 3: Electronic Signals and Reset Procedures.
	Part 4: Answer to Reset and Transmission Protocols.
	This standard defines a "close-coupled" card (up to 10 cm).
ISO 11568 (1994)	Banking: Key management (retail). The three parts of this standard define the principles, techniques, and life cycle of keys in a symmetric cipher scheme. Further parts are proposed to deal with key management in a public key environment.
ISO 14443-1 (2000)	Identification cards: Contactless integrated circuit(s) cards. Proximity cards. Part 1: Physical Characteristics. This specification covers the two most common contactless card types used today.

STANDARD NAME	TITLE AND DESCRIPTION
ISO 15693-2 (2000)	Identification cards: Contactless integrated circuit(s) cards. Vicinity cards. Part 2: Air Interface and Initialization. This standard covers cards for use at greater coupling distances than the other two contactless standards (up to 1.5m).
ITSO	The Integrated Transport Smartcard Organization (which was set up in the north of England but which is aiming for a European scope) has specified a smart card application for use across multiple public transport operators.
NPR 7402	The Netherlands National Chip Card Platform is designed to ensure compatibility across electronic purse, debit and credit card applications used in The Netherlands.
OpenCard	The Open Card Framework (OCF) specifications are designed to permit application and terminal development independent of the card design. See http://www.opencard.org.
PC/SC	The PC Smart Card architecture is an open architecture developed by a group of smart card and PC operating system vendors, notably CP8 Transac, HP, Microsoft, Schlumberger, and Siemens Nixdorf. It is intended to ensure interoperability between components from different vendors and across different hardware and software platforms. It comprises: Part 1: Introduction and Architecture Overview. Part 2: Interface Requirements for Compatible IC Cards and Readers. Part 3: Requirements for PC-Connected Interface Devices. Part 4: IFD Design Considerations and Reference Design Information. Part 5: ICC Resource Manager Definition. Part 6: ICC Service Provider Interface Definition. Part 7: Application Domain/Developer Design Considerations. Part 8: Recommendations for ICC Security and Privacy Devices. All of these documents can be downloaded from http://www.pcscworkgroup.com.

Standard Name	Title and Description
SET	Secure Electronic Transactions: A set of standards for credit card payment across networks, using the conventional cardholder-merchant-acquirer structure and the card number only. SET is a license-free open standard and can be downloaded from the Internet. Acquirers, merchants, and cardholders obtain their keys from and are authenticated by a certification server operated by the card scheme. SET is controlled by SETCo (www.setco.org).
WAP	Wireless Application Protocol: A set of 16 specifications published by the European Telecommunications Standards Institute on behalf of the WAP Forum, covering the content, transmission protocols, and security structures for transmission of Wireless Markup Language-encoded pages over mobile networks.

Appendix B: Glossary

Acceptor   The organization (usually a merchant) which accepts a card (e.g., in payment).

Acquirer   The bank which processes a merchant's transactions and passes them into the clearing system.

AID   Application identifier: the unique code associated with a card application, which allows the terminal to select a suitable application within the card for a given operation.

ALD   Application load certificate, used by Multos and similar technologies to authenticate an application being loaded to a card. The instance of an application being loaded is called an application load unit (ALU).

Antitearing   A card feature which protects the contents of memory if the card is removed before the end of the transaction.

Application   The program within a smart card which governs its external functions.

APDU   Application Protocol Data Unit: the term used in ISO 7816 for a message to or from a card application.

API   Application program interface: how a program communicates with, or uses, another program or service.

ASIC   Application-specific integrated circuit: a very large scale integrated circuit (a VLSI chip) designed for a specific customer and function (often on the basis of a programmable gate array).

ATC   Application transaction counter: a counter maintained within a chip card which increments by one for each transaction performed.

ATM   Automated teller machine (cash machine). Or, for data networks, asynchronous transfer mode.

ATR   Answer to reset: the data sent by a card to the reader when the card is first powered up.

Authentication   The process of verifying the identity and legitimacy of a person, object, or system.

Biometric   Identification of a person by a physical or behavioral characteristic (such as the way they sign their name, their fingerprint, or the marks on the iris of their eye).

CA   Certification authority: a body able to certify the identity of one or more parties to an exchange or transaction.

CAM   Card authentication method: the method (usually static or dynamic data authentication) used to verify that a card has come from a valid issuer and has not been tampered with.

CAT   Cardholder-activated terminal. Card schemes further subdivide CATs into groups, for example, low-value vending machines, limited-value (e.g., petrol pumps), and unlimited value on-line terminals (e.g., kiosks and ticket-booking systems).

Cardholder   The person to whom a personal card was issued (not necessarily the person holding the card).

CB   The French GIE Cartes Bancaires is an umbrella body controlling the card operations of the French banks. It sets standards as well as runs the data network. The current CB standard card does not conform with EMV, but a migration is planned.

CEN   Centre Européen pour la Normalisation (European Standards Center). *See also* EN.

CEPS   Common Electronic Purse Specifications: CEPS is a joint initiative involving many of the most important electronic purse schemes round the world (but not Mondex). CEPS specifies a core set of functions for which all CEPS-compatible purse should be able to exchange information. It sits on top of, rather than replacing, other electronic purse schemes.

Challenge-response   A form of authentication in which the system seeking authentication sends out a random "challenge." The object (e.g., the card or terminal) being authenticated performs a calculation on the challenge and responds with a result, from which the challenger can ascertain the authenticity or otherwise of the object. This method of authentication is much more secure than a simple password or other unvarying response.

Chip card   A card which embodies a "chip" (an integrated circuit). Also commonly known as a smart card, but the term "chip card" is often used to include those types of card which are not really "smart," such as memory cards.

Chinese remainder   A mathematical technique for performing modular arithmetic. It is used in smart cards for deriving digital signatures.

CLA   The class byte of an ISO command—see ISO 7816 part 4.

CLEF   Commercial Licensed Evaluation Facility: a body licensed to carry out security evaluations using the ITSEC criteria.

CMOS   Complementary metal-oxide silicon: a way of forming semiconductor material which uses less power than most other forms. *See also* HCMOS.

Combi card   A card which uses both contact and contactless technology.

Common Criteria   The Common Criteria for Information Security supersede and will eventually replace ITSEC and TCSEC (q.v.).

Contactless   Smart card technology using radio waves rather than contacts to energize and communicate with the chip inside the card.

CRC   Cyclic redundancy check: a check field often added to the end of a message, calculated as a polynomial from the rest of the message content. If a bit in the message is altered, then the CRC should alter.

Cryptogram   The result of a cryptographic operation.

Cryptology   The science of codes and ciphers (used in encryption).

Cryptoprocessor   A processor optimized for cryptographic functions (e.g., variable-length arithmetic, modular exponentiation, or DES encryption).

CVM   Cardholder verification method: the signature, password, PIN, or biometric used to check the identity of the cardholder, particularly for bank cards.

DDA   Dynamic data authentication: authentication of a card using a challenge-and-response mechanism.

DES   Data Encryption Standard (or Data Encryption Algorithm): the most widely used method for "symmetric" encryption (i.e., using the same key for encryption and decryption). The main source is ANSI X3.92.

DF   Dedicated file: the intermediate level of a card's file structure. DFs can hold data, EFs, or other DFs.

Diffie-Hellman   Diffie and Hellman were the first to describe viable public-key distribution and signature cryptograms in a paper in 1976. Their method, which is based on discrete logarithms, is still used in some systems, but RSA is more widely used in smart card schemes.

Digital cash   This term is applied to various schemes which represent money using electronic means. In the smart card world, value is usually stored on a card known as an electronic purse. Digital cash, however, normally consists of software "certificates" or tokens which can be stored on computer, or transferred to another party as payment.

Digital signature   An encrypted field, normally encrypted using the sender's private key, which is attached to a message to prove its source and integrity.

DSP   Digital signal processor: an integrated circuit or specialized computer for processing high-frequency analog signals.

EEPROM   Electrically erasable programmable read only memory: semiconductor memory which retains its memory without power, but can be changed at any time.

EF   Elementary file: the lowest level of a card's file structure. An EF may only contain data.

EFT-POS   Electronic funds transfer at point-of-sale: electronic payment.

Electronic purse   A card which stores value in the form of digital cash. An electronic purse is normally issued by a bank and the value it holds is the strict counterpart of legal tender. *See also* stored value card.

EMV   The Europay-Mastercard-Visa specifications for chip-based payment cards. EMV part 1 corresponds with (and generally conforms with) ISO 7816 parts 1–5; the other parts of this specification cover the details of a standard credit/debit application and the requirements for terminals.

EN   Euronorm or European Standard. Important ENs for smart cards include EN 726 (a multifunction telephone card) and EN 1546 (intersector electronic purse).

Encryption   Manipulating data to make it unreadable to anyone who does not possess the decryption key.

EPOS   Electronic point-of-sale (terminal): a networked and programmable electronic till.

E²PROM   *See* EEPROM.

ESD   Electrostatic discharge—the effect of discharging a high voltage but at a very low current, as when removing a woolen jumper or leaving a car after a long journey. ESD can be very harmful to electronic devices, particularly those using CMOS technology.

ETSI   European Telecommunications Standards Institute.

ETU   Elementary time unit: the "clock tick" on which all chip card timings are based.

Fabrication   The process of manufacturing the chip, which is used in a smart card.

FAR   False accept rate: the percentage of impostors accepted by a biometric or other identification check.

FERAM   Ferro-electric RAM: random access memory covered with an additional layer in a patented process to make it nonvolatile (i.e., it does not lose its memory when powered off). FERAM is much faster and uses less space than E^2PROM.

Flash memory   Semiconductor memory which can be written once, but can thereafter only be erased as a block. It is increasingly used for program storage, since it allows the program to be updated.

FPGA   Field programmable gate array: a semiconductor device which generates its outputs directly from its input states according to a "program" defined by the user.

FRAM   *See* FERAM.

FRR   False reject rate: the percentage of valid users rejected by a biometric or other identification check.

GSM   Global System for Mobile Communication: international standard for digital mobile telephony.

HCMOS   High-power CMOS: the technology used in most smart card microcontrollers.

HSM   Host security module (or hardware security module): a hardware device used for storing keys and performing cryptographic functions under control of a host computer.

IC   Integrated circuit.

IC card   Same as chip card. The banking industry prefers the term "IC card" or "ICC."

ID-1, ID-00   An ID-1 card is one having the format defined in ISO 7810. ID-00 is the alternative name for the "plug-in" form factor used in GSM SIMs and in SAMs.

IFD   Interface device: same as a card-accepting device or read-write unit, the equivalent of a card reader.

IMSI   International mobile subscriber identity: the ID of a GSM subscriber.

Integrity   (Of data or a message.) Not having been altered since it was originated.

ISO   International Standards Organization. The main ISO standard relating to smart cards is ISO 7816. Identification cards: integrated circuit cards with contacts. ISO 10536 and the draft standard 14443 cover, respectively, close-coupled and remotely coupled contactless cards. Many other standards covering aspects of security and computer systems operations are used by smart card systems.

ITSEC   Information Technology Security Evaluation Criteria: European standard for evaluating the security of commercial computer products. *See also* TCSEC.

ITU   International Telecommunications Union: the international body responsible for telecommunications coordination, the successor body to CCITT. *See also* ETSI.

Javacard   Application Program Interface specifications for running a subset of the Java language on a smart card. Java is an open, machine-independent language which offers a high level of protection between applications; it is thus well suited to a multiapplication smart card, although it imposes a higher overhead than conventional smart card operating systems.

Keys   In a modern encryption system, the algorithm is generally assumed to be known, and what is kept secret is the key. There are many different forms of key, each of which can be regarded as a string of meaningless bits until it is used to encode or decode a message.

Key escrow   One of the more emotive topics in cryptography is governments' desire to control the use of "strong" encryption, to prevent its use by criminals and enemies of the state. One method proposed to give this control, while still permitting the use of strong encryption, is key escrow: encryption users lodge a copy of their private keys with an accredited body, which agrees to surrender the keys to the government on production of a court order.

MAC   Message authentication code: a cryptographically derived block of data appended to a message to demonstrate that it has not been altered during transmission.

Mask   The fixed program of a microprocessor smart card.

MCPA   MasterCard Chip Payment Application: the scheme which governs chip card–based credit/debit transactions within the MasterCard system.

ME   Mobile equipment: the GSM name for a telephone or device used as a telephone.

MEL   Multos executable language: the intermediate code form in which Multos programs are loaded and executed.

Memory card   A chip card with memory, but controlled only by fixed logic rather than by a microprocessor.

MF   Master file: the top level of a card's file structure. A card always has a master file, which may contain data, DFs, or EFs.

Microprocessor   A semiconductor device which can execute a program. In a microprocessor-based smart card, the processor is combined with memory, power control, and other functions on a single "chip" of silicon.

Open Card (OCF)   The Open Card Framework is an architecture for cards and terminals intended to standardize the development of smart card and terminal applications. It is promoted by Apple, IBM, Netscape, NCI, and Sun, and is strongly linked with Java developments in the same area.

Personalization   Adding the individual card details to a card after manufacture. These will include the cardholder data in the chip's memory, usually the cardholder's name and an expiry date printed or embossed on the front. It may include other forms of personalization such as magnetic-stripe data or a photograph. During personalization, any variable program (in addition to the mask) may be stored in the card, as well as cryptographic keys.

PC/SC   The PC Smart Card architecture promoted by Microsoft and other smart card and PC operating system vendors, to standardize hardware and software interfaces for smart cards in PCs.

PGA   Programmable gate array. *See also* FPGA.

PIN   Personal identification number: a code (usually four to six digits) used as a password by a cardholder.

Pocket   In an electronic purse, a single store of value (e.g., one currency). A purse may have several pockets.

POS   Point of sale.

Public key   A public key encryption algorithm is one in which one key is published and the other kept secret.

PUK   PIN unblocking key (or personal unblocking key): a numeric code used to release a blocked application or card.

RAM   Random access memory (the equivalent of normal computer memory).

RFID   Radio frequency identification: a technology which allows an object or person to be identified at a distance, using radio waves to energize and communicate with some form of tag or card.

RISC   Reduced instruction set computer: a computer or microprocessor which, by operating with a smaller range of instructions, is able to achieve higher instruction speeds than conventional processors.

ROM   Read-only memory.

RPK   A new algorithm for public key encryption and authentication which operates at higher speeds than other algorithms.

RSA   The Rivest-Shamir-Adleman algorithm is the form of public key encryption most widely used today, particularly for digital signatures and key exchange.

SAM   Security application module: a chip normally used as part of a terminal to store keys and encryption algorithms securely. SAMs often use the same smart card technology as the associated cards, or a more specialized cryptographic chip.

SDA   Static data authentication: authentication of a card by means of a digitally signed copy of selected card data.

SET   Secure electronic transactions: protocols for securing Internet transactions.

SIM   Subscriber identity module: the personalization chip card in a GSM telephone.

SMS   Short message service: a form of transmission used in GSM telephony for short data messages.

Smart card   A card which incorporates a microprocessor chip and some form of storage. By extension, and in common usage, any form of chip card.

Windows for Smart Cards   Microsoft's multiapplication card platform.

Stored value card   A card which is used to store value such as loyalty points or credit for canteen meals. In Europe, the term is used to denote a card which is issued and redeemed within a closed circuit, in contrast with an electronic purse, which can be used to buy goods and services in the open market. In the United States, the term "stored value card" is used more widely, and can denote an electronic purse.

T=0/T=1   The asynchronous character and block protocols respectively defined by ISO 7816 part 3.

TASI   Terminal application services interface: the way that an application interfaces with the outside world (for use in testing an application or service).

TC   Transaction certificate: a value derived cryptographically from other transaction parameters, which enables the integrity and source of the transaction to be verified at a later date.

TCSEC   Trusted Computer Security Evaluation Criteria: the U.S. "Orange Book" requirements for evaluating the security of computer systems.

TTP   Trusted third party: an organization (usually government appointed or registered) which holds keys used for authentication purposes.

VIS   Visa IC card specification: the scheme which governs chip card-based credit/debit transactions within the Visa system.

VOP   Visa Open Platform: Visa's version of the Open Platform specifications, including some payment-specific functions and issuer controls.

WORM   Write once read many times (form of semiconductor memory).

Zero knowledge   A form of authentication in which the object demonstrates that it knows a secret, without disclosing that secret to the challenger (who may not know the secret). Most zero knowledge tests make use of public key cryptography, where the secret represents the private key or a function thereof. *See also* "challenge-response."

Appendix C:
Bibliography (Smart Card
Security References)

The widest range of references on cryptography generally is to be found on the various Internet newsgroups which deal with this subject. They include sci.crypt, comp.security, and many subsidiary newsgroups under these. Sci.crypt includes a frequently asked questions (FAQ) file which can be downloaded and which includes a long list of references including surveys, history, and analytical work. Alt.security and alt.hacker frequently contain references to successful and attempted smart card attacks.

There are several "link farms" with large numbers of links relating to cryptography and computer security. For example, see

www.securityserver.com or www.cs.hut.fi/crypto/

Internet searches using "smart card" + "security" normally yield useful results. It is advisable to restrict the search using further application, geographical, or date criteria.

Other references of specific interest to smart card security include:

Anderson, R. J., "Why Cryptosystems Fail," *Communications of the ACM*, Vol. 37, No. 11, 1994, pp. 32–40.

Anderson, R. J., and S. J. Bezuidenhout, "Cryptographic Credit Control in Pre-Payment Metering Systems," *IEEE Symposium on Security and Privacy*, 1995, pp. 15–23.

Anderson, R. J., and M. Kuhn, "Tamper Resistance—a Cautionary Note," *Proceedings of the Second Workshop on Electronic Commerce*, USENIX Association, Oakland, CA, 1996.

Bank for International Settlements, *Security of Electronic Money*, Basle, Switzerland, 1996.

Bank for International Settlements, *Electronic Money: Consumer Protection, Law Enforcement, Supervisory and Cross Border Issues*, Basle, Switzerland, 1997.

Beller, M. J., and Y. Yacobi, "Fully-Fledged Two-Way Public Key Authentication and Key Agreement for Low-Cost Terminals," *Electronics Letters (UK)*, Vol. 29, No. 11, pp. 999–1001.

Burmester, M., Y. Desmedt, and T. Beth, "Efficient Zero-Knowledge Identification Schemes for Smart Cards," *Computer Journal*, Vol. 35, No. 1, 1992, pp. 21–29.

Carlson, R., "Utilizing Smart Cards and PKI Technology to Solve E-Business Security Issues," corporate publication, Authentic8, Doncaster (VIC, Australia), 1999.

Frank, J. N., "Smart Cards Meet Biometrics," *Card Technology*, Sept./Oct. 1996, pp. 30–38.

Garfinkel, S., and G. Spafford, *Web Security & Commerce*, Sebastopol, CA: O'Reilly & Associates, Inc., 1997.

Hendry, M., *Practical Computer Network Security*, Norwood, MA: Artech House, 1995.

Hendry, M., "Security Is More Than a Card Game," *Smart Card '97 Conference Proceedings*, QMS, Peterborough, England, 1997.

Holloway, C., "The IBM Personal Security Card," *Smart Card '91 Conference Proceedings*, Agestream, Peterborough, England, 1991.

Janson, P., and M. Waidner, "Electronic Payment Systems," *SI Informatik/Informatique*, Switzerland, Vol. 3, 1995, pp. 10–15.

Konigs, H.-P., "Cryptographic Identification Methods for Smart Cards in the Process of Standardization," *IEEE Communications*, Vol. 29, No. 6, pp. 42–48.

Lim, C. H., et al., "Smart Card Reader," *IEEE Transactions on Consumer Electronics*, Vol. 39, No. 1, pp. 6–12.

Mondex International, "Mondex: Security by Design," corporate publication, London, 1995.

Rhee, M. Y., *Cryptography and Secure Communications*, Singapore: McGraw-Hill, 1994.

Schaumüller-Bichl, I., "Card Security: An Overview," *Smart Card 2000 Conference Proceedings*, North Holland, Amsterdam, 1991, pp. 19–27.

Schneier, B., *Applied Cryptography: Protocols, Algorithms and Source Code in C*, New York: Wiley, 1996.

Schneier, B., *Secrets and Lies: Digital Security in a Networked World*, New York: Wiley, 2000.

Sociedade Interbancaria de Servicos, SA, "Multibanco Electronic Purse— Description of Scheme," corporate publication, Lisbon, Portugal, 1993.

Thomasson, J.-P., "Advances in Smartcard IC Technology," SGS-Thomson technical article TA164, Paris, France, 1996.

Van Renesse, R. L., *Optical Document Security*, Norwood, MA: Artech House, 1997.

Waidner, M., "Secure Billing and Payment over the Internet," *Rapperswil Networking Forum*, Rapperswil, Switzerland, 1995.

About the Author

Mike Hendry is an independent consultant in payment systems and electronic commerce, with an emphasis on smart cards and transaction security. He has an M.A. in engineering from the University of Cambridge, UK, and an M.B.A. from IMI Geneva. He is multilingual and works with retailers, banks, and service providers throughout Europe on technology and business strategy issues.

Mr. Hendry has been closely involved with the implementation of chip-based credit, debit, and purse cards in the United Kingdom and in Europe. He has worked with both card issuers and major retailers on the specification of their business, operational and technical requirements, and the longer-term opportunities afforded by the technology. His current activities include product and systems design for e-commerce and mobile phone "top-ups," as well as cross-border acquiring for credit and debit card transactions.

Index

For further information on these and other Artech House titles, including previously considered out-of-print books now available through our In-Print-Forever® (IPF®) program, contact:

Artech House
685 Canton Street
Norwood, MA 02062
Phone: 781-769-9750
Fax: 781-769-6334
e-mail: artech@artechhouse.com

Artech House
46 Gillingham Street
London SW1V 1AH UK
Phone: +44 (0)20 7596-8750
Fax: +44 (0)20 7630-0166
e-mail: artech-uk@artechhouse.com

Find us on the World Wide Web at:
www.artechhouse.com